An Introduction to Modeling and Simulation of Particulate Flows

COMPUTATIONAL SCIENCE & ENGINEERING

Computational Science and Engineering (CS&E) is widely accepted, along with theory and experiment, as a crucial third mode of scientific investigation and engineering design. This series publishes research monographs, advanced undergraduate- and graduate-level textbooks, and other volumes of interest to a wide segment of the community of computational scientists and engineers. The series also includes volumes addressed to users of CS&E methods by targeting specific groups of professionals whose work relies extensively on computational science and engineering.

Series Volumes

Zohdi, T. I., *An Introduction to Modeling and Simulation of Particulate Flows*

Biegler, Lorenz T., Omar Ghattas, Matthias Heinkenschloss, David Keyes, and Bart van Bloemen Waanders, Editors, *Real-Time PDE-Constrained Optimization*

Chen, Zhangxin, Guanren Huan, and Yuanle Ma, *Computational Methods for Multiphase Flows in Porous Media*

Shapira, Yair, *Solving PDEs in C++: Numerical Methods in a Unified Object-Oriented Approach*

An Introduction to Modeling and Simulation of Particulate Flows

T. I. Zohdi
University of California–Berkeley
Berkeley, California

siam
Society for Industrial and Applied Mathematics
Philadelphia

10 9 8 7 6 5 4 3 2 1

The cover was produced from images created by and used with permission of the Scientific Computing and Imaging (SCI) Institute, University of Utah; J. Bielak, D. O'Hallaron, L. Ramirez-Guzman, and T. Tu, Carnegie Mellon University; O. Ghattas, University of Texas at Austin; K. Ma and H. Yu, University of California, Davis; and Mark R. Petersen, Los Alamos National Laboratory. More information about the images is available at *http://www.siam.org/books/series/csecover/php*.

Library of Congress Cataloging-in-Publication Data

Zohdi, Tarek I.
An introduction to modeling and simulation of particulate flows / Tarek I. Zohdi.
p. cm. -- (Computational science and engineering)
ISBN 978-0-89871 6-27-6 (alk. paper)
1. Granular materials--Fluid dynamics--Mathematical models. I. Title.

TA418.78.Z64 2007
620'.43--dc22

2007061728

Dedicated to my patient wife, Britta,
and my mother and father, Omnia and Magd

Contents

List of Figures

Preface

The study of "granular" or "particulate" media is wide ranging. Classical examples include the study of natural materials, such as sand and gravel, associated with coastal erosion, landslides, and avalanches. A concise introduction is given by Duran [61]. Many manufactured materials also fall within this class of problems.[1] For general overviews of granular media, we refer the reader to Jaeger and Nagel [100], [101], Nagel [151], Liu et al. [139], Liu and Nagel [140], Jaeger and Nagel [102], Jaeger et al. [103]–[105], and Jaeger and Nagel [106]; the extensive works of Hutter and collaborators: Tai et al. [188]–[190], Gray et al. [80], Wieland et al. [201], Berezin et al. [28], Gray and Hutter [81], Gray [82], Hutter [96], Hutter et al. [97], Hutter and Rajagopal [98], Koch et al. [126], Greve and Hutter [85], and Hutter et al. [99]; the works of Behringer and collaborators: Behringer [22], Behringer and Baxter [21], Behringer and Miller [23], and Behringer et al. [24]; the works of Jenkins and collaborators: Jenkins and Strack [107], Jenkins and La Ragione [108], Jenkins and Koenders [109], and Jenkins et al. [110]; and the works of Torquato and collaborators: Torquato [194], Kansaal et al. [119], and Donev et al. [55]–[59]. In this monograph, we focus on a subset of the very large field of granular materials, namely, *fluidized (lower-density) particulate flows.*[2]

Recently, several modern applications, primarily driven by microtechnology, have emerged where a successful analysis requires the simulation of flowing particulate media involving simultaneous near-field interaction between charged particles and momentum exchange through mechanical contact.[3] For example, in many systems containing flowing particles below the one millimeter scale, the particles can acquire relatively large electrostatic charges, leading to significant interparticle near-field forces. In some cases, the

[1] Over half (by weight) of the raw materials handled in chemical industries appear in granulated or particulate form. The resulting structural properties of solid products which originate as granulated or particulate media, and which are transported and constructed using flow processes, are outside the scope of this monograph. For more details, see, for example, Aboudi [1], Hashin [90], Mura [150], Nemat-Nasser and Hori [152], Torquato [194], and Zohdi and Wriggers [216].

[2] It is worth noting that fast computational methods, in particular efficient contact search techniques, for the treatment of densely packed granular or particulate media, in the absence of near-field forces, can be found in the recent work of Pöschel and Schwager [167]. Such techniques are outside the scope of the present work, but they are relatively easy to implement.

[3] For example, industrial processes such as chemical mechanical planarization (CMP), which involves using chemically reacting particles embedded in fluid (gas or liquid) to ablate rough small-scale surfaces flat, have become important in the success of many micro- and nanotechnologies. For a review of CMP practice and applications, see Luo and Dornfeld [143]–[146].

near-field forces could be due to magnetic effects, or they could be purposely induced.[4] Charged material can lead to inconsistent "clean" manufacturing processes, for example, due to difficulties with dust control, although intentional charging of particulate material can be quite useful in some applications, for example, in electrostatic copiers, inkjet printers, and powder coating machines. The presence of near-field interaction forces can produce particulate flows that are significantly different from purely contact-driven scenarios. Determining the dynamics of such materials is important in accurately describing the flow of powders, which form the basis of microfabrication. Near-field forces can lead to particle clustering, resulting in inconsistent fabrication quality. Therefore, neglecting such near-field effects can lead to a gross miscalculation of the characteristics of such flows.[5] Thus, an issue of overriding importance to the successful characterization of such flows is the development of models and reliable computational techniques to simulate the dynamics of multibody particulate systems involving near-field interaction and contact simultaneously (including thermal effects). This is the primary focus of this monograph.

A central objective of this work is to provide basic models and numerical solution strategies for the direct simulation of flowing particulate media that can be achieved within a relatively standard desktop or laptop computing environment. A primary assumption is that the objects in the flow are considered to be small enough to be idealized as particles and that the effects of their rotation with respect to their mass centers is unimportant to their overall motion.[6] Our primary concern is with particulate media that are "fluidized," i.e., they are not densely packed together. Oftentimes, such media are referred to as "granular gases." In particular, the initial chapters of the monograph are dedicated to so-called dry particulate flows, where there is no significant interstitial fluid. *However, while this monograph focuses almost exclusively on the dry problem, Chapter* 8 *gives an introduction to strongly coupled (surrounding) fluid/particle interaction scenarios.* Also, an introduction to computational optical techniques for particulate media is provided. Simulations described in upcoming chapters can be found at `http://www.siam.org/books/cs04`.

Ideally, in an attempt to reduce laboratory expenses, one would like to make predictions of a complex particulate flow's behavior by numerical simulations, with the primary goal being to minimize time-consuming trial and error experiments. The recent dramatic increase in computational power available for mathematical modeling and simulation raises the possibility that modern numerical methods can play a significant role in the analysis of complex particulate flows. This fact has motivated the work presented in this monograph. This work can be viewed as a research monograph, suitable for use in a first-year graduate course for students in the applied sciences, engineering, and applied mathemat-

[4]For many engineering materials, some surface adhesion persists even when no explicit charging has occurred. For example, see Tabor [186] or Johnson [111].

[5]For example, on the atomic scale, forces of attraction can arise from a temporary dipole created by fluctuating electron distributions around an atom. This will induce a dipole on a neighboring atom, and if the induced dipole is directed in the same way as the first atom, the two molecules associated with these atoms will attract one another. Between two atoms, such a force acts over a nanometer; however, when two small-scale (1–100 microns) particles approach one another, the effect is greatly multiplied and the forces act over much larger distances. Also, for example, repulsion forces can arise due to ionization of the particle surfaces or due to the adsorption of ions onto the surfaces of particles. The combination of attraction and repulsion forces is called a near-field force. It is worth noting that near-field forces can be introduced into a model in order to mimic much smaller scale effects attributed to chemical potentials, interstitial fluid, etc., which do not necessarily have as their basis a "charge."

[6]However, even in the event that the particles are not extremely small, we assume that any "spin" of the particles is small enough to neglect lift forces that may arise from the interaction with the surrounding fluid.

ics with an interest in the computational analysis of complex particulate flows. Although it is tempting, a survey of all possible modeling and computational techniques will not be undertaken, since the field is growing at an extremely rapid rate. *This monograph is designed to provide a basic introduction, using models that are as simple as possible.* Finally, I am certain that, despite painstaking efforts, there remain errors of one sort or another. Therefore, if readers find such errata, I would appreciate if they would contact me at `zohdi@newton.berkeley.edu`.

T. I. Zohdi

Berkeley, CA
November 2006

Chapter 1
Fundamentals

When the dimensions of a body are insignificant to the description of its motion or the action of forces on it, the body may be idealized as a particle, i.e., a piece of material occupying a point in space and perhaps moving as time passes. In the next few sections, we briefly review some essential concepts that will be needed later in the analysis of particles.

1.1 Notation

In this work, boldface symbols imply vectors or tensors. *A fixed Cartesian coordinate system will be used throughout.* The unit vectors for such a system are given by the mutually orthogonal triad $(\boldsymbol{e}_1, \boldsymbol{e}_2, \boldsymbol{e}_3)$. For the inner product of two vectors $\boldsymbol{u}$ and $\boldsymbol{v}$, we have in three dimensions

$$\boldsymbol{u} \cdot \boldsymbol{v} = \sum_{i=1}^{3} v_i u_i = u_1 v_1 + u_2 v_2 + u_3 v_3 = ||\boldsymbol{u}|| ||\boldsymbol{v}|| \cos\theta, \tag{1.1}$$

where

$$||\boldsymbol{u}|| = \sqrt{u_1^2 + u_2^2 + u_3^2} \tag{1.2}$$

represents the Euclidean norm in R^3 and θ is the angle between the two vectors. We recall that a norm has three main characteristics for any two bounded vectors $\boldsymbol{u}$ and $\boldsymbol{v}$ ($||\boldsymbol{u}|| < \infty$ and $||\boldsymbol{v}|| < \infty$):

- $||\boldsymbol{u}|| > 0$, and $||\boldsymbol{u}|| = 0$ if and only if $\boldsymbol{u} = \boldsymbol{0}$,
- $||\boldsymbol{u} + \boldsymbol{v}|| \leq ||\boldsymbol{u}|| + ||\boldsymbol{v}||$, and
- $||\gamma \boldsymbol{u}|| \leq |\gamma| ||\boldsymbol{u}||$, where γ is a scalar.

Two vectors are said to be orthogonal if $\boldsymbol{u} \cdot \boldsymbol{v} = 0$. The cross (vector) product of two vectors is

$$\boldsymbol{u} \times \boldsymbol{v} = -\boldsymbol{v} \times \boldsymbol{u} = \begin{vmatrix} e_1 & e_2 & e_3 \\ u_1 & u_2 & u_3 \\ v_1 & v_2 & v_3 \end{vmatrix} = ||\boldsymbol{u}|| ||\boldsymbol{v}|| \sin\theta \, \boldsymbol{n}, \tag{1.3}$$

where $\boldsymbol{n}$ is the unit normal to the plane formed by the vectors $\boldsymbol{u}$ and $\boldsymbol{v}$.

The temporal differentiation of a vector is given by

$$\frac{d}{dt}\boldsymbol{u}(t) = \frac{du_1(t)}{dt}\boldsymbol{e}_1 + \frac{du_2(t)}{dt}\boldsymbol{e}_2 + \frac{du_3(t)}{dt}\boldsymbol{e}_3 = \dot{u}_1\boldsymbol{e}_1 + \dot{u}_2\boldsymbol{e}_2 + \dot{u}_3\boldsymbol{e}_3. \tag{1.4}$$

The spatial gradient of a scalar (a dilation to a vector) is given by

$$\nabla\phi = \left(\boldsymbol{e}_1\frac{\partial\phi}{\partial x_1} + \boldsymbol{e}_2\frac{\partial\phi}{\partial x_2} + \boldsymbol{e}_3\frac{\partial\phi}{\partial x_3}\right). \tag{1.5}$$

The gradient of a vector is a direct extension of the preceding definition. For example, $\nabla\boldsymbol{u}$ has components of $\frac{\partial u_i}{\partial x_j}$. The divergence of a vector (a contraction to a scalar) is defined by

$$\nabla\cdot\boldsymbol{u} = \left(\boldsymbol{e}_1\frac{\partial}{\partial x_1} + \boldsymbol{e}_2\frac{\partial}{\partial x_2} + \boldsymbol{e}_3\frac{\partial}{\partial x_3}\right)\cdot(u_1\boldsymbol{e}_1 + u_2\boldsymbol{e}_2 + u_3\boldsymbol{e}_3) = \left(\frac{\partial u_1}{\partial x_1} + \frac{\partial u_2}{\partial x_2} + \frac{\partial u_3}{\partial x_3}\right). \tag{1.6}$$

The curl of a vector is defined as

$$\nabla\times\boldsymbol{u} = \begin{vmatrix} e_1 & e_2 & e_3 \\ \frac{\partial}{\partial x_1} & \frac{\partial}{\partial x_2} & \frac{\partial}{\partial x_3} \\ u_1 & u_2 & u_3 \end{vmatrix}. \tag{1.7}$$

1.2 Kinematics of a single particle

We denote the position of a point in space by the vector $\boldsymbol{r}$. The instantaneous velocity of a point is given by the limit

$$\boldsymbol{v} = \lim_{\Delta t\to 0}\frac{\boldsymbol{r}(t+\Delta t) - \boldsymbol{r}(t)}{\Delta t} = \frac{d\boldsymbol{r}}{dt} = \dot{\boldsymbol{r}}. \tag{1.8}$$

The instantaneous acceleration of a point is given by the limit

$$\boldsymbol{a} = \lim_{\Delta t\to 0}\frac{\boldsymbol{v}(t+\Delta t) - \boldsymbol{v}(t)}{\Delta t} = \frac{d\boldsymbol{v}}{dt} = \dot{\boldsymbol{v}} = \ddot{\boldsymbol{r}}. \tag{1.9}$$

In fixed Cartesian coordinates, we have

$$\boldsymbol{r} = r_1\boldsymbol{e}_1 + r_2\boldsymbol{e}_2 + r_3\boldsymbol{e}_3, \tag{1.10}$$

$$\boldsymbol{v} = \dot{\boldsymbol{r}} = \dot{r}_1\boldsymbol{e}_1 + \dot{r}_2\boldsymbol{e}_2 + \dot{r}_3\boldsymbol{e}_3, \tag{1.11}$$

and

$$\boldsymbol{a} = \ddot{\boldsymbol{r}} = \ddot{r}_1\boldsymbol{e}_1 + \ddot{r}_2\boldsymbol{e}_2 + \ddot{r}_3\boldsymbol{e}_3. \tag{1.12}$$

Their magnitudes are denoted by $||\boldsymbol{r}|| = \sqrt{\boldsymbol{r}\cdot\boldsymbol{r}}$, $||\boldsymbol{v}|| = \sqrt{\boldsymbol{v}\cdot\boldsymbol{v}}$, and $||\boldsymbol{a}|| = \sqrt{\boldsymbol{a}\cdot\boldsymbol{a}}$. The relative motion of a point i with respect to a point j is denoted by $\boldsymbol{r}_{i-j} = \boldsymbol{r}_i - \boldsymbol{r}_j$, $\boldsymbol{v}_{i-j} = \boldsymbol{v}_i - \boldsymbol{v}_j$, and $\boldsymbol{a}_{i-j} = \boldsymbol{a}_i - \boldsymbol{a}_j$.

1.3 Kinetics of a single particle

Throughout this monograph, the fundamental relation between force and acceleration is given by Newton's second law of motion, in vector form:

$$\boldsymbol{\Psi} = m\boldsymbol{a}, \tag{1.13}$$

where $\boldsymbol{\Psi}$ is the sum (resultant) of all the applied forces acting on mass m.

1.3.1 Work, energy, and power

A closely related concept is that of work and energy. The differential amount of work done by a force acting through a differential displacement is

$$dW = \boldsymbol{\Psi} \cdot d\boldsymbol{r}. \tag{1.14}$$

Therefore, the total amount of work performed by a force over a displacement history is

$$W_{1\to 2} = \int_{\boldsymbol{r}(t_1)}^{\boldsymbol{r}(t_2)} \boldsymbol{\Psi} \cdot d\boldsymbol{r} = \int_{\boldsymbol{r}(t_1)}^{\boldsymbol{r}(t_2)} m\boldsymbol{a} \cdot d\boldsymbol{r} = \int_{\boldsymbol{r}(t_1)}^{\boldsymbol{r}(t_2)} m\boldsymbol{v} \cdot d\boldsymbol{v} = \frac{1}{2}m(\boldsymbol{v}_2 \cdot \boldsymbol{v}_2 - \boldsymbol{v}_1 \cdot \boldsymbol{v}_1) \stackrel{\text{def}}{=} T_2 - T_1, \tag{1.15}$$

where $T \stackrel{\text{def}}{=} \frac{1}{2}m\boldsymbol{v} \cdot \boldsymbol{v}$ is known as the kinetic energy.[7] Therefore, we may write

$$T_1 + W_{1\to 2} = T_2. \tag{1.16}$$

If the forces can be written in the form

$$dV = -\boldsymbol{\Psi} \cdot d\boldsymbol{r}, \tag{1.17}$$

then

$$W_{1\to 2} = -\int_{\boldsymbol{r}(t_1)}^{\boldsymbol{r}(t_2)} dV = V(\boldsymbol{r}(t_1)) - V(\boldsymbol{r}(t_2)), \tag{1.18}$$

where

$$\boldsymbol{\Psi} = -\nabla V. \tag{1.19}$$

Such a force is said to be conservative. Furthermore, it is easy to show that a conservative force must satisfy

$$\nabla \times \boldsymbol{\Psi} = \boldsymbol{0}. \tag{1.20}$$

The work done by a conservative force on any closed path is zero, since

$$-\int_{\boldsymbol{r}(t_1)}^{\boldsymbol{r}(t_2)} dV = V(\boldsymbol{r}(t_1)) - V(\boldsymbol{r}(t_2)) = \int_{\boldsymbol{r}(t_2)}^{\boldsymbol{r}(t_1)} dV \Rightarrow \int_{\boldsymbol{r}(t_1)}^{\boldsymbol{r}(t_2)} dV + \int_{\boldsymbol{r}(t_2)}^{\boldsymbol{r}(t_1)} dV = 0. \tag{1.21}$$

As a consequence, for a conservative system,

$$T_1 + V_1 = T_2 + V_2. \tag{1.22}$$

Also, power can be defined as the time rate of change of work:

$$\frac{dW}{dt} = \frac{\boldsymbol{\Psi} \cdot d\boldsymbol{r}}{dt} = \boldsymbol{\Psi} \cdot \boldsymbol{v}. \tag{1.23}$$

[7] The chain rule was used to write $\boldsymbol{a} \cdot d\boldsymbol{r} = \boldsymbol{v} \cdot d\boldsymbol{v}$.

1.3.2 Properties of a potential

As we have indicated, a force field $\boldsymbol{\Psi}$ is said to be conservative if and only if there exists a continuously differentiable scalar field V such that $\boldsymbol{\Psi} = -\nabla V$. Therefore, a necessary and sufficient condition for a particle to be in equilibrium is that

$$\boldsymbol{\Psi} = -\nabla V = \mathbf{0}. \tag{1.24}$$

In other words,

$$\frac{\partial V}{\partial x_1} = 0, \quad \frac{\partial V}{\partial x_2} = 0, \quad \text{and} \quad \frac{\partial V}{\partial x_3} = 0. \tag{1.25}$$

Forces acting on a particle (1) that are always directed toward or away from another point and (2) whose magnitude depends only on the distance between the particle and the point of attraction/repulsion are called *central forces.* They have the form

$$\boldsymbol{\Psi} = -\mathcal{C}(||\boldsymbol{r} - \boldsymbol{r}_o||)\frac{\boldsymbol{r} - \boldsymbol{r}_o}{||\boldsymbol{r} - \boldsymbol{r}_o||} = \mathcal{C}(||\boldsymbol{r} - \boldsymbol{r}_o||)\boldsymbol{n}, \tag{1.26}$$

where $\boldsymbol{r}$ is the position of the particle, $\boldsymbol{r}_o$ is the position of a point that the particle is attracted toward or repulsed from, and

$$\boldsymbol{n} = \frac{\boldsymbol{r}_o - \boldsymbol{r}}{||\boldsymbol{r} - \boldsymbol{r}_o||}. \tag{1.27}$$

The central force is one of attraction if

$$\mathcal{C}(||\boldsymbol{r} - \boldsymbol{r}_o||) > 0 \tag{1.28}$$

and one of repulsion if

$$\mathcal{C}(||\boldsymbol{r} - \boldsymbol{r}_o||) < 0. \tag{1.29}$$

We remark that a central force field is always conservative, since $\nabla \times \boldsymbol{\Psi} = \mathbf{0}$. Now consider the specific choice

$$V = \underbrace{\frac{\alpha_1||\boldsymbol{r} - \boldsymbol{r}_o||^{-\beta_1+1}}{-\beta_1 + 1}}_{\text{attraction}} - \underbrace{\frac{\alpha_2||\boldsymbol{r} - \boldsymbol{r}_o||^{-\beta_2+1}}{-\beta_2 + 1}}_{\text{repulsion}}, \tag{1.30}$$

where all of the parameters, the α's and β's, are nonnegative. The gradient yields

$$-\nabla V = \boldsymbol{\Psi} = \left(\alpha_1||\boldsymbol{r} - \boldsymbol{r}_o||^{-\beta_1} - \alpha_2||\boldsymbol{r} - \boldsymbol{r}_o||^{-\beta_2}\right)\boldsymbol{n}, \tag{1.31}$$

which is repeatedly used later in this monograph. If a particle which is displaced slightly from an equilibrium point tends to return to that point, then we call that point a point of stability or stable point, and the equilibrium is said to be stable. Otherwise, we say that the point is one of instability and the equilibrium is unstable. *A necessary and sufficient condition for a point of equilibrium to be stable is that the potential V at that point be a minimum.* The general condition by which a potential is stable for the multidimensional case can be determined by studying the properties of the Hessian of V,

$$[\mathbb{H}] \stackrel{\text{def}}{=} \begin{bmatrix} \frac{\partial^2 V}{\partial x_1 \partial x_1} & \frac{\partial^2 V}{\partial x_1 \partial x_2} & \frac{\partial^2 V}{\partial x_1 \partial x_3} \\ \frac{\partial^2 V}{\partial x_2 \partial x_1} & \frac{\partial^2 V}{\partial x_2 \partial x_2} & \frac{\partial^2 V}{\partial x_2 \partial x_3} \\ \frac{\partial^2 V}{\partial x_3 \partial x_1} & \frac{\partial^2 V}{\partial x_3 \partial x_2} & \frac{\partial^2 V}{\partial x_3 \partial x_3} \end{bmatrix}, \tag{1.32}$$

around an equilibrium point. A sufficient condition for V to attain a minimum at an equilibrium point is for the Hessian to be positive definite (which implies that V is locally convex). For more details, see Hale and Kocak [88].

Remark. Provided that the α's and β's are selected appropriately, the chosen central force potential form is stable for motion in the normal direction, i.e., the line connecting the centers of particles in particle-particle interaction.[8] In order to determine stable parameter combinations, consider a potential function for a single particle, in one-dimensional motion, representing the motion in the normal direction, attracted to and repulsed from a point r_o, measured by the coordinate r,

$$V = \frac{\alpha_1}{-\beta_1+1}|r-r_o|^{-\beta_1+1} - \frac{\alpha_2}{-\beta_2+1}|r-r_o|^{-\beta_2+1}, \tag{1.33}$$

whose derivative produces the form of interaction forces introduced earlier:

$$\Psi = -\frac{\partial V}{\partial r} = \left(\alpha_1|r-r_o|^{-\beta_1} - \alpha_2|r-r_o|^{-\beta_2}\right) n, \tag{1.34}$$

where $n = \frac{r_o - r}{|r-r_o|}$. For stability, we require

$$\frac{\partial^2 V}{\partial r^2} = -\alpha_1\beta_1|r-r_o|^{-\beta_1-1} + \alpha_2\beta_2|r-r_o|^{-\beta_2-1} > 0. \tag{1.35}$$

A static equilibrium point, $r = r_e$, can be calculated from $\Psi(|r_e - r_o|) = -\alpha_1|r_e - r_o|^{-\beta_1} + \alpha_2|r_e - r_o|^{-\beta_2} = 0$, which implies

$$|r_e - r_o| = \left(\frac{\alpha_2}{\alpha_1}\right)^{\frac{1}{-\beta_1+\beta_2}}. \tag{1.36}$$

Inserting Equation (1.36) into Equation (1.35) yields a restriction for stability

$$\frac{\beta_2}{\beta_1} > 1. \tag{1.37}$$

Thus, for the appropriate choices of the α's and β's, the central force potential in Equation (1.30) is stable for motion in the normal direction, i.e., the line connecting the centers of the particles. For disturbances in directions orthogonal to the normal direction, the potential is neutrally stable, i.e., the Hessian's determinant is zero, thus indicating that the potential does not change for such perturbations.

1.3.3 Impulse and momentum

Newton's second law can be rewritten as

$$\boldsymbol{\Psi} = \frac{d(m\boldsymbol{v})}{dt} \Rightarrow \boldsymbol{G}(t_1) + \int_{t_1}^{t_2} \boldsymbol{\Psi}\, dt = \boldsymbol{G}(t_2), \tag{1.38}$$

[8]For disturbances in directions orthogonal to the normal direction, the potential is neutrally stable, i.e., the Hessian's determinant is zero, thus indicating that the potential does not change for such perturbations. The motion analysis in the normal direction is relevant for central forces of the type under consideration.

where

$$G(t_1) = (m\boldsymbol{v})|_{t=t_1} \tag{1.39}$$

is the linear momentum. Clearly, if

$$\boldsymbol{\Psi} = \boldsymbol{0}, \tag{1.40}$$

then

$$G(t_1) = G(t_2), \tag{1.41}$$

and linear momentum is said to be conserved.

A related quantity is the angular momentum. About the origin,

$$\boldsymbol{H}_o \stackrel{\text{def}}{=} \boldsymbol{r} \times m\boldsymbol{v}. \tag{1.42}$$

Clearly, the moment $\mathcal{M}$ implies

$$\mathcal{M} = \boldsymbol{r} \times \boldsymbol{\Psi} = \frac{d(\boldsymbol{r} \times m\boldsymbol{v})}{dt} \Rightarrow \boldsymbol{H}_o(t_1) + \int_{t_1}^{t_2} \underbrace{\boldsymbol{r} \times \boldsymbol{\Psi}}_{\mathcal{M}}\, dt = \boldsymbol{H}_o(t_2). \tag{1.43}$$

Thus, if

$$\mathcal{M} = \boldsymbol{0}, \tag{1.44}$$

then

$$\boldsymbol{H}_o(t_1) = \boldsymbol{H}_o(t_2), \tag{1.45}$$

and angular momentum is said to be conserved.

1.4 Systems of particles

We now discuss the dynamics of a system of N_p particles. Let $\boldsymbol{r}_i$, $i = 1, 2, 3, \ldots, N_p$, be the position vectors of a system of particles.

1.4.1 Linear momentum

The position vector of the center of mass of the system is given by

$$\boldsymbol{r}_{cm} \stackrel{\text{def}}{=} \frac{\sum_{i=1}^{N_p} m_i \boldsymbol{r}_i}{\sum_{i=1}^{N_p} m_i} = \frac{1}{M}\sum_{i=1}^{N_p} m_i \boldsymbol{r}_i. \tag{1.46}$$

Consider a decomposition of the position vector for particle i of the form

$$\boldsymbol{r}_i = \boldsymbol{r}_{cm} + \boldsymbol{r}_{i-cm}. \tag{1.47}$$

The linear momentum of a system of particles is given by

$$\sum_{i=1}^{N_p} \underbrace{m_i \dot{\boldsymbol{r}}_i}_{\boldsymbol{G}_i} = \sum_{i=1}^{N_p} m_i (\dot{\boldsymbol{r}}_{cm} + \dot{\boldsymbol{r}}_{i-cm}) = \sum_{i=1}^{N_p} m_i \dot{\boldsymbol{r}}_{cm} = \dot{\boldsymbol{r}}_{cm} \sum_{i=1}^{N_p} m_i \stackrel{\text{def}}{=} \boldsymbol{G}_{cm}, \tag{1.48}$$

since

$$\sum_{i=1}^{N_p} m_i \dot{\boldsymbol{r}}_{i-cm} = \boldsymbol{0}. \tag{1.49}$$

Thus, the linear momentum of any system with constant mass is the product of the mass and the velocity of its center of mass; furthermore,

$$\dot{\boldsymbol{G}}_{cm} = M \ddot{\boldsymbol{r}}_{cm}. \tag{1.50}$$

When considering a system of particles, it is advantageous to decompose the forces acting on a particle into forces from external sources and those from internal sources:

$$\boldsymbol{\Psi} = \boldsymbol{\Psi}^{EXT} + \boldsymbol{\Psi}^{INT}. \tag{1.51}$$

Summing over all particles in the system leads to cancellation of the internal forces. For example, consider the external forces $\boldsymbol{\Psi}_i^{EXT}$ and internal forces $\boldsymbol{\Psi}_i^{INT}$ acting on a single member of the system of particles. Newton's second law states

$$m_i \ddot{\boldsymbol{r}}_i = \boldsymbol{\Psi}_i^{EXT} + \boldsymbol{\Psi}_i^{INT}. \tag{1.52}$$

Now sum over all the particles in the system to obtain

$$\sum_{i=1}^{N_p} m_i \ddot{\boldsymbol{r}}_i = M \ddot{\boldsymbol{r}}_{cm} = \sum_{i=1}^{N_p} (\boldsymbol{\Psi}_i^{EXT} + \boldsymbol{\Psi}_i^{INT}) = \sum_{i=1}^{N_p} \boldsymbol{\Psi}_i^{EXT} + \underbrace{\sum_{i=1}^{N_p} \boldsymbol{\Psi}_i^{INT}}_{=\boldsymbol{0}} = \sum_{i=1}^{N_p} \boldsymbol{\Psi}_i^{EXT}, \tag{1.53}$$

since the internal forces in the system are equal in magnitude and opposite in direction. Thus,

$$\dot{\boldsymbol{G}}_{cm} = M \ddot{\boldsymbol{r}}_{cm} = \sum_{i=1}^{N_p} \boldsymbol{\Psi}_i^{EXT}. \tag{1.54}$$

Thus, the impulse-momentum relation reads

$$\boldsymbol{G}_{cm}(t_1) + \sum_{i=1}^{N_p} \int_{t_1}^{t_2} \boldsymbol{\Psi}_i^{EXT}\, dt = \boldsymbol{G}_{cm}(t_2). \tag{1.55}$$

1.4.2 Energy principles

The work-energy principle for many particles is formally the same as that for a single particle:

$$\sum_{i=1}^{N_p} T_{i,1} + \sum_{i=1}^{N_p} W_{i,1\to 2} = \sum_{i=1}^{N_p} T_{i,2}, \tag{1.56}$$

where $\sum W_{i,1\to 2}$ represents all of the work done by the external and internal forces. It is advantageous to decompose the kinetic energy into the translation of the center of mass and the motion relative to the center of mass. This is achieved by writing

$$\boldsymbol{v}_i = \boldsymbol{v}_{cm} + \dot{\boldsymbol{r}}_{i-cm}, \tag{1.57}$$

which yields

$$\sum_{i=1}^{N_p} T_i = \sum_{i=1}^{N_p} \frac{1}{2} m_i (\boldsymbol{v}_{cm} + \dot{\boldsymbol{r}}_{i-cm}) \cdot (\boldsymbol{v}_{cm} + \dot{\boldsymbol{r}}_{i-cm})$$
$$= \sum_{i=1}^{N_p} \frac{1}{2} m_i \boldsymbol{v}_{cm} \cdot \boldsymbol{v}_{cm} + \sum_{i=1}^{N_p} \frac{1}{2} m_i \dot{\boldsymbol{r}}_{i-cm} \cdot \dot{\boldsymbol{r}}_{i-cm}. \tag{1.58}$$

If the entire system is rigid, the second term takes on the meaning of rotation around the center of mass.

1.4.3 Remarks on scaling

Historically, when experimental testing of a physically enormous or minute true-scale system was either impossible or prohibitively expensive, one scaled up (or down) the system size and tested a model of manageable dimensions. A key to comparing a model of normalized dimensions to that of the true model is the concept of dynamic similitude and dimensionless parameters. Similarly, in order to illustrate generic computational methods without having to tie them to a specific application, we frequently use a fixed control volume of normalized dimensions. Therefore, it is important to be able to determine the correlation between the parameters for the normalized model and a true system that has different dimensions. This is achieved by similitude. A few basic concepts are important:

- *Geometric similarity* requires that the two models be of the same shape and that all linear dimensions of the models be related by a constant scale factor.

- *Kinematic similarity* of two models requires the velocities at corresponding points to be in the same direction and to be related by a constant scale factor.

- When two models have force distributions such that identical types of forces are parallel and are related in magnitude by a constant scale factor at all corresponding points, the models are said to be *dynamically similar*, i.e., they exhibit similitude. The requirements for dynamic similarity are the most restrictive: *two models must possess both geometric and kinematic similarity to be dynamically similar. In other words, geometric and kinematic similarity are necessary for dynamic similarity.*

A standard approach to determining the conditions under which two models are similar is to normalize the governing differential equations and boundary conditions. Similitude may be present when two physical phenomena are governed by identical differential equations and boundary conditions. Similitude is obtained when governing equations and boundary conditions have the same dimensionless form. This is obtained by duplicating the dimensionless coefficients that appear in the normalization of the models.

For example, consider the governing equation for a particle i within a system of particles ($j \neq i$):

$$m_i \ddot{\boldsymbol{r}}_i = \sum_{j \neq i}^{N_p} \left(\alpha_{1ij} ||\boldsymbol{r}_i - \boldsymbol{r}_j||^{-\beta_1} - \alpha_{2ij} ||\boldsymbol{r}_i - \boldsymbol{r}_j||^{-\beta_2} \right) \boldsymbol{n}_{ij}, \tag{1.59}$$

where the normal direction is determined by the difference in the position vectors of the particles' centers:

$$\boldsymbol{n}_{ij} \stackrel{\text{def}}{=} \frac{\boldsymbol{r}_j - \boldsymbol{r}_i}{||\boldsymbol{r}_i - \boldsymbol{r}_j||}. \tag{1.60}$$

In order to perform the normalization of the model in Equation (1.59), we introduce the following dimensionless parameters:

- $\boldsymbol{r}^* \stackrel{\text{def}}{=} \frac{\boldsymbol{r}}{L}$,
- $t^* \stackrel{\text{def}}{=} \frac{t}{T}$.

The quantities that appear in Equation (1.59) become

- $m_i \ddot{\boldsymbol{r}}_i = m_i \frac{L}{T^2} \frac{d^2 \boldsymbol{r}_i^*}{dt^{*2}}$,
- $\alpha_{1ij} ||\boldsymbol{r}_i - \boldsymbol{r}_j||^{-\beta_1} = \alpha_{1ij} L^{-\beta_1} ||\boldsymbol{r}_i^* - \boldsymbol{r}_j^*||^{-\beta_1}$,
- $\alpha_{2ij} ||\boldsymbol{r}_i - \boldsymbol{r}_j||^{-\beta_2} = \alpha_{2ij} L^{-\beta_2} ||\boldsymbol{r}_i^* - \boldsymbol{r}_j^*||^{-\beta_2}$,

where $\boldsymbol{n}_{ij}$ remains unchanged. Substituting these relations into Equation (1.59) yields

$$\frac{d^2 \boldsymbol{r}_i^*}{dt^{*2}} = \sum_{j \neq i}^{N_p} \left(\frac{\alpha_{1ij}}{m_i} T^2 L^{-(\beta_1+1)} ||\boldsymbol{r}_i^* - \boldsymbol{r}_j^*||^{-\beta_1} - \frac{\alpha_{2ij}}{m_i} T^2 L^{-(\beta_2+1)} ||\boldsymbol{r}_i^* - \boldsymbol{r}_j^*||^{-\beta_2} \right) \boldsymbol{n}_{ij}. \tag{1.61}$$

Thus, two dimensionless parameters, which must be the same for two systems to exhibit similitude between one another, are

- $\frac{\alpha_{1ij}}{m_i} T^2 L^{-(\beta_1+1)}$,
- $\frac{\alpha_{2ij}}{m_i} T^2 L^{-(\beta_2+1)}$.

In other words,

$$\left(\frac{\alpha_{1ij}}{m_i} T^2 L^{-(\beta_1+1)} \right)_{system\ 1} = \left(\frac{\alpha_{1ij}}{m_i} T^2 L^{-(\beta_1+1)} \right)_{system\ 2} \tag{1.62}$$

and

$$\left(\frac{\alpha_{2ij}}{m_i} T^2 L^{-(\beta_2+1)} \right)_{system\ 1} = \left(\frac{\alpha_{2ij}}{m_i} T^2 L^{-(\beta_2+1)} \right)_{system\ 2} \tag{1.63}$$

must hold simultaneously for the models to produce comparable results.

Chapter 2

Modeling of particulate flows

As indicated in the preface, in this introductory monograph the objects in the flow are assumed to be small enough to be considered (idealized) as particles, spherical in shape, and the effects of their rotation with respect to their mass center are assumed unimportant to their overall motion.

2.1 Particulate flow in the presence of near-fields

We consider a group of nonintersecting particles (N_p in total).[9] The equation of motion for the ith particle in a flow is

$$m_i \ddot{\boldsymbol{r}}_i = \boldsymbol{\Psi}_i^{tot}(\boldsymbol{r}_1, \boldsymbol{r}_2, \ldots, \boldsymbol{r}_{N_p}), \tag{2.1}$$

where $\boldsymbol{r}_i$ is the position vector of the ith particle and $\boldsymbol{\Psi}_i^{tot}$ represents all forces acting on particle i. Specifically,

$$\boldsymbol{\Psi}_i^{tot} = \boldsymbol{\Psi}_i^{nf} + \boldsymbol{\Psi}_i^{con} + \boldsymbol{\Psi}_i^{fric} \tag{2.2}$$

represents the sum of forces due to near-field interaction ($\boldsymbol{\Psi}^{nf}$), normal contact forces ($\boldsymbol{\Psi}^{con}$), and friction ($\boldsymbol{\Psi}^{fric}$). We consider the following relatively general central-force attraction-repulsion form for the near-field forces induced by all particles on particle i:

$$\boldsymbol{\Psi}_i^{nf} = \sum_{j \neq i}^{N_p} \left(\underbrace{\alpha_{1ij} ||\boldsymbol{r}_i - \boldsymbol{r}_j||^{-\beta_1}}_{\text{attraction}} - \underbrace{\alpha_{2ij} ||\boldsymbol{r}_i - \boldsymbol{r}_j||^{-\beta_2}}_{\text{repulsion}} \right) \underbrace{\boldsymbol{n}_{ij}}_{\text{unit vector}}, \tag{2.3}$$

where $|| \cdot ||$ represents the Euclidean norm in R^3, the α's and β's are nonnegative, and the normal direction is determined by the difference in the position vectors of the particles' centers

$$\boldsymbol{n}_{ij} \stackrel{\text{def}}{=} \frac{\boldsymbol{r}_j - \boldsymbol{r}_i}{||\boldsymbol{r}_i - \boldsymbol{r}_j||}. \tag{2.4}$$

---

[9] The approach in this chapter draws from methods developed in Zohdi [212] and [217].

Figure 2.1. *Compression and recovery of two impacting particles (Zohdi* [212]*).*

Remark. Later in the analysis, it is convenient to employ the following (per unit mass[2]) decompositions for the key near-field parameters for the force imparted on particle i by particle j, and vice versa:[10]

- $\alpha_{1ij} = \overline{\alpha}_1 m_i m_j$,
- $\alpha_{2ij} = \overline{\alpha}_2 m_i m_j$.

2.2 Mechanical contact with near-field interaction

We now consider cases where mechanical contact occurs between particles in the presence of near-field interaction. A primary simplifying assumption is made: *the particles remain spherical after impact, i.e., any permanent deformation is considered negligible.* For two colliding particles i and j, normal to the line of impact, a balance of linear momentum relating the states before impact (time $= t$) and after impact (time $= t + \delta t$) reads as

$$m_i v_{in}(t) + m_j v_{jn}(t) + \int_t^{t+\delta t} \boldsymbol{E}_i \cdot \boldsymbol{n}_{ij}\, dt + \int_t^{t+\delta t} \boldsymbol{E}_j \cdot \boldsymbol{n}_{ij}\, dt = m_i v_{in}(t+\delta t) + m_j v_{jn}(t+\delta t), \tag{2.5}$$

where the subscript n denotes the normal component of the velocity (along the line connecting particle centers) and the $\boldsymbol{E}$'s represent all forces induced by near-field interaction with other particles, as well as all other external forces, if any, applied to the pair. If one isolates one of the members of the colliding pair, then

$$m_i v_{in}(t) + \int_t^{t+\delta t} I_n\, dt + \int_t^{t+\delta t} \boldsymbol{E}_i \cdot \boldsymbol{n}_{ij}\, dt = m_i v_{in}(t + \delta t), \tag{2.6}$$

where $\int_t^{t+\delta t} I_n\, dt$ is the total normal impulse due to impact. For a pair of particles undergoing impact, let us consider a decomposition of the collision event (Figure 2.1) into a compression (δt_1) and a recovery (δt_2) phase, i.e., $\delta t = \delta t_1 + \delta t_2$. Between the compression and recovery

[10] Alternatively, if the near-fields are related to the amount of surface area, this scaling could be done per unit area.

phases, the particles achieve a common velocity,[11] denoted by v_{cn}, at the intermediate time $t+\delta t_1$. We may write for particle i, along the normal, in the compression phase of impact,

$$m_i v_{in}(t) + \int_t^{t+\delta t_1} I_n\,dt + \int_t^{t+\delta t_1} \boldsymbol{E}_i \cdot \boldsymbol{n}_{ij}\,dt = m_i v_{cn}, \tag{2.7}$$

and, in the recovery phase,

$$m_i v_{cn} + \int_{t+\delta t_1}^{t+\delta t} I_n\,dt + \int_{t+\delta t_1}^{t+\delta t} \boldsymbol{E}_i \cdot \boldsymbol{n}_{ij}\,dt = m_i v_{in}(t+\delta t). \tag{2.8}$$

For the other particle (j), in the compression phase,

$$m_j v_{jn}(t) - \int_t^{t+\delta t_1} I_n\,dt + \int_t^{t+\delta t_1} \boldsymbol{E}_j \cdot \boldsymbol{n}_{ij}\,dt = m_j v_{cn}, \tag{2.9}$$

and, in the recovery phase,

$$m_j v_{cn} - \int_{t+\delta t_1}^{t+\delta t} I_n\,dt + \int_{t+\delta t_1}^{t+\delta t} \boldsymbol{E}_j \cdot \boldsymbol{n}_{ij}\,dt = m_j v_{jn}(t+\delta t). \tag{2.10}$$

This leads to an expression for the coefficient of restitution:

$$\begin{aligned} e \stackrel{\text{def}}{=} \frac{\int_{t+\delta t_1}^{t+\delta t} I_n\,dt}{\int_t^{t+\delta t_1} I_n\,dt} &= \frac{m_i(v_{in}(t+\delta t) - v_{cn}) - E_{in}(t+\delta t_1, t+\delta t)}{m_i(v_{cn} - v_{in}(t)) - E_{in}(t, t+\delta t_1)} \\ &= \frac{-m_j(v_{jn}(t+\delta t) - v_{cn}) + E_{jn}(t+\delta t_1, t+\delta t)}{-m_j(v_{cn} - v_{jn}(t)) + E_{jn}(t, t+\delta t_1)}, \end{aligned} \tag{2.11}$$

where

$$\begin{aligned} E_{in}(t+\delta t_1, t+\delta t) &\stackrel{\text{def}}{=} \int_{t+\delta t_1}^{t+\delta t} \boldsymbol{E}_i \cdot \boldsymbol{n}_{ij}\,dt, \\ E_{jn}(t+\delta t_1, t+\delta t) &\stackrel{\text{def}}{=} \int_{t+\delta t_1}^{t+\delta t} \boldsymbol{E}_j \cdot \boldsymbol{n}_{ij}\,dt, \\ E_{in}(t, t+\delta t_1) &\stackrel{\text{def}}{=} \int_t^{t+\delta t_1} \boldsymbol{E}_i \cdot \boldsymbol{n}_{ij}\,dt, \\ E_{jn}(t, t+\delta t_1) &\stackrel{\text{def}}{=} \int_t^{t+\delta t_1} \boldsymbol{E}_j \cdot \boldsymbol{n}_{ij}\,dt. \end{aligned} \tag{2.12}$$

If we eliminate v_{cn}, we obtain an expression for e:

$$e = \frac{v_{jn}(t+\delta t) - v_{in}(t+\delta t) + \Delta_{ij}(t+\delta t_1, t+\delta t)}{v_{in}(t) - v_{jn}(t) + \Delta_{ij}(t, t+\delta t_1)}, \tag{2.13}$$

[11] A common normal velocity for particles should be interpreted as indicating that the relative velocity in the normal direction between particle centers is zero.

where[12]

$$\Delta_{ij}(t+\delta t_1, t+\delta t) \stackrel{\text{def}}{=} \frac{1}{m_i}E_{in}(t+\delta t_1, t+\delta t) - \frac{1}{m_j}E_{jn}(t+\delta t_1, t+\delta t) \tag{2.14}$$

and

$$\Delta_{ij}(t, t+\delta t_1) \stackrel{\text{def}}{=} \frac{1}{m_i}E_{in}(t, t+\delta t_1) - \frac{1}{m_j}E_{jn}(t, t+\delta t_1). \tag{2.15}$$

Thus, we may rewrite Equation (2.13) as

$$v_{jn}(t+\delta t) = v_{in}(t+\delta t) - \Delta_{ij}(t+\delta t_1, t+\delta t) + e\left(v_{in}(t) - v_{jn}(t) + \Delta_{ij}(t, t+\delta t_1)\right). \tag{2.16}$$

It is convenient to denote the average force acting on the particle from external sources as

$$\overline{E}_{in} \stackrel{\text{def}}{=} \frac{1}{\delta t}\int_t^{t+\delta t} \boldsymbol{E}_i \cdot \boldsymbol{n}_{ij}\, dt. \tag{2.17}$$

If e is explicitly known, then, combining Equations (2.13) and (2.5), one can write

$$\begin{aligned} v_{in}(t+\delta t) &= \frac{m_i v_{in}(t) + m_j(v_{jn}(t) - e(v_{in}(t) - v_{jn}(t)))}{m_i + m_j} \\ &+ \frac{(\overline{E}_{in} + \overline{E}_{jn})\delta t - m_j(e\Delta_{ij}(t, t+\delta t_1) - \Delta_{ij}(t+\delta t_1, t+\delta t))}{m_i + m_j}, \end{aligned} \tag{2.18}$$

and, once $\boldsymbol{v}_{in}(t+\delta t)$ is known, one can subsequently also solve for $v_{jn}(t+\delta t)$ via Equation (2.16).

Remark. Later, it will be useful to define the average impulsive normal contact force between the particles acting during the impact event as

$$\overline{I}_n \stackrel{\text{def}}{=} \frac{1}{\delta t}\int_t^{t+\delta t} I_n\, dt = \frac{m_i(v_{in}(t+\delta t) - v_{in}(t))}{\delta t} - \overline{E}_{in}. \tag{2.19}$$

In particular, as will be done later in the analysis, when we discretize the equations of motion with a discrete (finite difference) time step of Δt, where $\delta t \ll \Delta t$, we shall define the impulsive normal contact contribution to the total force acting on a particle, $\boldsymbol{\Psi}_i^{tot} = \boldsymbol{\Psi}_i^{nf} + \boldsymbol{\Psi}_i^{con} + \boldsymbol{\Psi}_i^{fric}$ (Equation (2.2)), to be

$$\boldsymbol{\Psi}^{con} = \frac{\overline{I}_n \delta t}{\Delta t}\boldsymbol{n}_{ij}. \tag{2.20}$$

Furthermore, at the implementation level, we choose $\delta t = \gamma \Delta t$, where $0 < \gamma \ll 1$ and Δt is the time step discretization size, which will be introduced later in the work.[13] We assume $\delta t_1 + \delta t_2 = \delta t_1 + e\delta t_1$, which immediately allows the definitions

$$\delta t_1 = \frac{\gamma \Delta t}{1+e} \quad \text{and} \quad \delta t_2 = \frac{e\gamma\Delta t}{1+e}. \tag{2.21}$$

[12]This collapses to the classical expression for the ratio of the relative velocities before and after impact if the near-field forces are negligible:

$$e \stackrel{\text{def}}{=} \frac{v_{jn}(t+\delta t) - v_{in}(t+\delta t)}{v_{in}(t) - v_{jn}(t)}.$$

[13]A typical choice is $0 < \gamma \le 0.01$. Typically, the system is insensitive to γ below 0.01.

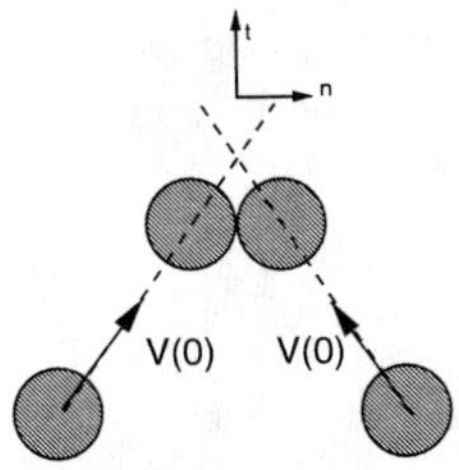

Figure 2.2. *Two identical particles approaching one another (Zohdi* [212]*).*

These results are consistent with the fact that the recovery time vanishes (all compression and no recovery) for $e \to 0$ (asymptotically "plastic") and, as $e \to 1$, the recovery time equals the compression time ($\delta t_2 = \delta t_1$, asymptotically "elastic"). If $e = 1$, there is no loss in energy, while if $e = 0$, there is a maximum loss in energy. For a more detailed analysis of impact duration times, see Johnson [111].

Remark. It is obvious that for a deeper understanding of the fields within a particle, it must be treated as a deformable continuum. This will inevitably require the spatial discretization, for example, using the finite element method (FEM), of the body (particle). The implementation, theory, and application of FEM is the subject of an immense literature. For general references on the subject, see the well-known books of Bathe [18], Becker et al. [19], Hughes [95], Szabo and Babúska [185], and Zienkiewicz and Taylor [207]. For work specifically focusing on the continuum mechanics of particles, see Zohdi and Wriggers [216]. For a detailed numerical analysis of multifield interaction between bodies, see Wriggers [203].

2.3 Kinetic energy dissipation

Consider two identical particles approaching one another (Figure 2.2) in the absence of near-field interaction. One can directly write for the kinetic energy (T), before and after impact,

$$T(t + \delta t) - T(t) = T(t)(e^2 - 1) \leq 0, \tag{2.22}$$

thus indicating the rather obvious fact that energy is lost with each subsequent impact for $e < 1$. Now consider a group of flowing particles, each with different velocity. We may decompose the velocity of each particle by defining

$$\boldsymbol{v}_{cm} = \frac{1}{M} \sum_{i=1}^{N_p} m_i \boldsymbol{v}_i \tag{2.23}$$

and $M = \sum_{i=1}^{N_p} m_i$, leading to

$$\boldsymbol{v}_i(t) = \boldsymbol{v}_{cm}(t) + \delta \boldsymbol{v}_i(t), \tag{2.24}$$

where $\boldsymbol{v}_{cm}(t)$ is the mean velocity of the group of particles and $\delta\boldsymbol{v}_i(t)$ is a purely fluctuating (about the mean) part of the velocity. For the entire group of particles at time $= t$,

$$\sum_{i=1}^{N_p} m_i\boldsymbol{v}_i(t)\cdot\boldsymbol{v}_i(t) = \sum_{i=1}^{N_p} m_i(\boldsymbol{v}_{cm}(t)+\delta\boldsymbol{v}_i(t))\cdot(\boldsymbol{v}_{cm}(t)+\delta\boldsymbol{v}_i(t))$$
$$= M\boldsymbol{v}_{cm}(t)\cdot\boldsymbol{v}_{cm}(t) + 2\boldsymbol{v}_{cm}(t)\cdot\underbrace{\sum_{i=1}^{N_p} m_i\delta\boldsymbol{v}_i(t)}_{=\boldsymbol{0}} + \sum_{i=1}^{N_p} m_i\delta\boldsymbol{v}_i(t)\cdot\delta\boldsymbol{v}_i(t). \tag{2.25}$$

For any later stage, the mean velocity ($\boldsymbol{v}_{cm}$) remains constant, and we have

$$\sum_{i=1}^{N_p} m_i(\boldsymbol{v}_i(t+\delta t)\cdot\boldsymbol{v}_i(t+\delta t)) = M\boldsymbol{v}_{cm}(t)\cdot\boldsymbol{v}_{cm}(t) + \sum_{i=1}^{N_p} m_i\delta\boldsymbol{v}_i(t+\delta t)\cdot\delta\boldsymbol{v}_i(t+\delta t). \tag{2.26}$$

Subtracting Equation (2.25) from Equation (2.26) yields

$$\sum_{i=1}^{N_p} m_i\boldsymbol{v}_i(t+\delta t)\cdot\boldsymbol{v}_i(t+\delta t) - \sum_{i=1}^{N_p} m_i\boldsymbol{v}_i(t)\cdot\boldsymbol{v}_i(t)$$
$$= \sum_{i=1}^{N_p} m_i\delta\boldsymbol{v}_i(t+\delta t)\cdot\delta\boldsymbol{v}_i(t+\delta t) - \sum_{i=1}^{N_p} m_i\delta\boldsymbol{v}_i(t)\cdot\delta\boldsymbol{v}_i(t)$$
$$\geq e^2\sum_{i=1}^{N_p} m_i\delta\boldsymbol{v}_i(t)\cdot\delta\boldsymbol{v}_i(t) - \sum_{i=1}^{N_p} m_i\delta\boldsymbol{v}_i(t)\cdot\delta\boldsymbol{v}_i(t)$$
$$= (e^2-1)\sum_{i=1}^{N_p} m_i\delta\boldsymbol{v}_i(t)\cdot\delta\boldsymbol{v}_i(t),$$
$$\geq (e^2-1)\sum_{i=1}^{N_p} m_i\boldsymbol{v}_i(t)\cdot\boldsymbol{v}_i(t), \tag{2.27}$$

where the first inequality arises because not all particles will experience an impact from one stage to the next and the second inequality arises because the perturbation's energy (that associated with $\delta\boldsymbol{v}$) must be smaller than the total (that associated with $\boldsymbol{v}$). Thus, *in the absence of near-field interaction*, we should expect

$$e^2-1 \leq \frac{T(t+\delta t)-T(t)}{T(t)} \leq 0. \tag{2.28}$$

Remark. In order to help characterize the overall behavior of the motion, it is advantageous to decompose the kinetic energy per unit mass into the bulk motion of the center of mass and the motion relative to the center of mass:

$$\overline{T}(t) = \frac{T(t)}{M} = \underbrace{\frac{1}{2}\boldsymbol{v}_{cm}(t)\cdot\boldsymbol{v}_{cm}(t)}_{\stackrel{\text{def}}{=}T_b = \text{bulk motion energy}} + \underbrace{\frac{1}{2M}\sum_{i=1}^{N_p} m_i\delta\boldsymbol{v}_i(t)\cdot\delta\boldsymbol{v}_i(t)}_{\stackrel{\text{def}}{=}T_r = \text{relative motion energy}}. \tag{2.29}$$

Clearly, the identification of the "bulk" and "relative" parts is important in some applications, and this decomposition provides a natural way of characterizing the particulate flow.[14] We note that the system momentum is conserved provided there are no external forces applied to the entire system. For values of $e < 1$, the relative motion will eventually "die out" *if no near-field forces are present.*

Remark. Sometimes expressions of the form

$$\sum_{i=1}^{N_p} m_i \boldsymbol{v}_i \cdot \boldsymbol{v}_i - M\boldsymbol{v}_{cm} \cdot \boldsymbol{v}_{cm} = \sum_{i=1}^{N_p} m_i \delta\boldsymbol{v}_i \cdot \delta\boldsymbol{v}_i \tag{2.30}$$

are termed "granular gas temperatures."

2.4 Incorporating friction

To incorporate frictional stick-slip phenomena during impact, for a general particle pair (i and j), the tangential velocities at the beginning of the impact time interval (time $= t$) are computed by subtracting the relative normal velocity from the total relative velocity:

$$\boldsymbol{v}_{jt}(t) - \boldsymbol{v}_{it}(t) = (\boldsymbol{v}_j(t) - \boldsymbol{v}_i(t)) - \big((\boldsymbol{v}_j(t) - \boldsymbol{v}_i(t)) \cdot \boldsymbol{n}_{ij}\big)\boldsymbol{n}_{ij}. \tag{2.31}$$

One then writes the equation for tangential momentum change during impact for the ith particle:

$$m_i v_{it}(t) - \overline{I}_f \delta t + \overline{E}_{it}\delta t = m_i v_{ct}, \tag{2.32}$$

where the friction contribution is

$$\overline{I}_f = \frac{1}{\delta t}\int_t^{t+\delta t} I_f \, dt, \tag{2.33}$$

the total contribution from all other particles in the tangential direction ($\boldsymbol{\tau}_{ij}$) is

$$\overline{E}_{it} = \frac{1}{\delta t}\int_t^{t+\delta t} \boldsymbol{E}_i \cdot \boldsymbol{\tau}_{ij} \, dt, \tag{2.34}$$

and v_{ct} is the common velocity of particles i and j in the tangential direction.[15] Similarly, for the jth particle we have

$$m_j v_{jt}(t) + \overline{I}_f \delta t + \overline{E}_{jt}\delta t = m_j v_{ct}. \tag{2.35}$$

There are two unknowns, $\overline{I}_f$ and v_{ct}. The main quantity of interest is $\overline{I}_f$, which can be solved for as

$$\overline{I}_f = \frac{\left(\frac{\overline{E}_{it}}{m_i} - \frac{\overline{E}_{jt}}{m_j}\right)\delta t + v_{it}(t) - v_{jt}(t)}{\left(\frac{1}{m_i} + \frac{1}{m_j}\right)\delta t}. \tag{2.36}$$

---

[14]An example is mixing processes.

[15]They do not move relative to one another.

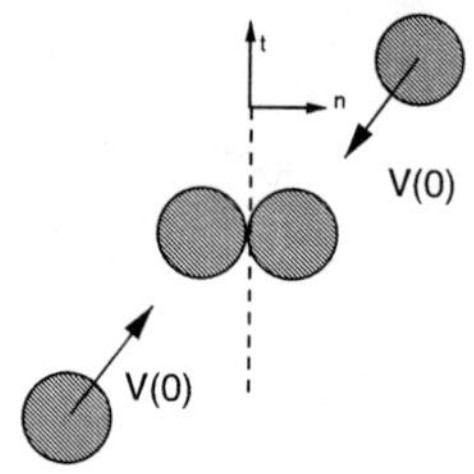

Figure 2.3. *Two identical particles approaching one another (Zohdi* [212]*).*

Thus, consistent with stick-slip models of Coulomb friction, one first assumes that no slip occurs. If

$$|\overline{I}_f| > \mu_s |\overline{I}_n|, \tag{2.37}$$

where

$$\mu_s \geq \mu_d \tag{2.38}$$

is the coefficient of *static* friction, then slip must occur and a dynamic sliding friction model is used. If sliding occurs, the friction force is assumed to be proportional to the normal force and opposite to the direction of relative tangential motion, i.e.,

$$\boldsymbol{\Psi}_i^{fric} \stackrel{\text{def}}{=} \mu_d ||\boldsymbol{\Psi}^{con}|| \frac{\boldsymbol{v}_{jt} - \boldsymbol{v}_{it}}{||\boldsymbol{v}_{jt} - \boldsymbol{v}_{it}||} = -\boldsymbol{\Psi}_j^{fric}. \tag{2.39}$$

2.4.1 Limitations on friction coefficients

There are limitations on the friction coefficients for such models to make physical sense. For example, reconsider the simple case of two identical particles (Figure 2.3), in the absence of near-field forces, approaching one another with velocity $\boldsymbol{v}(t)$, which can be decomposed into normal and tangential components:

$$\boldsymbol{v}(t) = v_n(t)\boldsymbol{e}_n + v_\tau(t)\boldsymbol{e}_\tau. \tag{2.40}$$

Now consider the pre- and postimpact kinetic energy, which is identical for each of the particles, assuming sliding (dynamic friction):

$$T(t) = \frac{1}{2} m(v_n^2(t) + v_\tau^2(t)) \tag{2.41}$$

and

$$T(t + \delta t) = \frac{1}{2} m(v_n^2(t + \delta t) + v_\tau^2(t + \delta t)). \tag{2.42}$$

Assuming sliding takes place, for either particle, the impulse-momentum relation can be written as

$$m v_n(t) + \int_t^{t+\delta t} I_n \, dt = m v_n(t + \delta t) \tag{2.43}$$

in the normal direction and

$$mv_t(t) - \int_t^{t+\delta t} \mu_d I_n \, dt = mv_t(t+\delta t) \tag{2.44}$$

in the tangential direction. For the normal direction,

$$\int_t^{t+\delta t} I_n \, dt = m(v_n(t+\delta t) - v_n(t)) = -(1+e)mv_n(t). \tag{2.45}$$

Substituting this relation into the conservation of momentum relation in the tangential direction, we have

$$v_\tau(t+\delta t) = v_\tau(t) - (1+e)v_n(t)\mu_d. \tag{2.46}$$

Now consider the restriction that the friction forces cannot be so large that they reverse the initial tangential motion. Mathematically, this restriction can be written as

$$v_\tau(t+\delta t) = v_\tau(t) - (1+e)v_n(t)\mu_d \geq 0, \tag{2.47}$$

which leads to the expression

$$\mu_d \leq \frac{v_t(t)}{v_n(t)(1+e)}. \tag{2.48}$$

Thus, the dynamic coefficient of friction must be restricted in order to make physical sense. Qualitatively, as e grows the restrictions on the coefficients of friction are more severe, although the author has determined that, typically, values of $\mu_d \leq 0.5$ are usually acceptable for the applications considered. For more general analyses of the validity of mechanical models involving friction, see, for example, Oden and Pires [154], Martins and Oden [147], Kikuchi and Oden [123], Klarbring [125], Tuzun and Walton [196], or Cho and Barber [42].

Remark. One can determine the coefficient of friction that maximizes energy loss by substituting Equation (2.46) into (2.42) and computing

$$\frac{\partial T(t+\delta t)}{\partial \mu_d} = 0 \Rightarrow \mu_d^* = \frac{v_t(t)}{v_n(t)(1+e)}, \tag{2.49}$$

which is the maximum value of μ_d dictated by Equation (2.48).[16]

2.4.2 Velocity-dependent coefficients of restitution

It is important to realize that, in reality, the phenomenological parameter e depends on the severity of the impact velocity. For extensive experimental data, see Goldsmith [79], or see Johnson [111] for a more detailed analytical treatment. Qualitatively, the coefficient of restitution has behavior as shown in Figure 2.4. A mathematical idealization of the behavior can be constructed as

$$e \stackrel{\text{def}}{=} \max\left(e_o\left(1 - \frac{\Delta v_n}{v^*}\right), e^-\right), \tag{2.50}$$

[16] The second derivative indicates $\frac{\partial^2 T(t+\delta t)}{\partial \mu_d^2} > 0$, so μ_d^* is a minimizer of $T(t+\delta t)$. This result, which is intuitive, implies that increasing the sliding friction coefficients allows more energy to be dissipated.

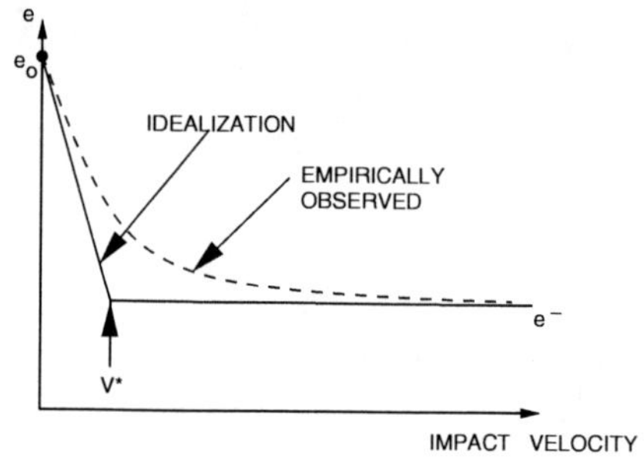

Figure 2.4. *Qualitative behavior of the coefficient of restitution with impact velocity (Zohdi* [212]*).*

where v^* is a critical threshold velocity (normalization) parameter, the relative velocity of approach is defined by

$$\Delta v_n \stackrel{\text{def}}{=} |v_{jn}(t) - v_{in}(t)|, \tag{2.51}$$

and e^- is a lower limit to the coefficient of restitution.

Chapter 3

Iterative solution schemes

3.1 Simple temporal discretization

Generally, methods for the time integration of differential equations fall within two broad categories: (1) implicit and (2) explicit. In order to clearly distinguish between the two approaches, we study a generic equation of the form

$$\dot{r} = \mathcal{G}(r, t). \tag{3.1}$$

If we discretize the differential equation,

$$\dot{r} \approx \frac{r(t + \Delta t) - r(t)}{\Delta t} \approx \mathcal{G}(r, t). \tag{3.2}$$

A primary question is, at which time should we evaluate the equation? If we use time $= t$, then

$$\dot{r}|_t = \frac{r(t + \Delta t) - r(t)}{\Delta t} = \mathcal{G}(r(t), t) \Rightarrow r(t + \Delta t) = r(t) + \Delta t \mathcal{G}(r(t), t), \tag{3.3}$$

which yields an *explicit* expression for $r(t + \Delta t)$. This is often referred to as a *forward Euler* scheme. If we use time $= t + \Delta t$, then

$$\dot{r}|_{t+\Delta t} = \frac{r(t + \Delta t) - r(t)}{\Delta t} = \mathcal{G}(r(t + \Delta t), t + \Delta t), \tag{3.4}$$

and therefore

$$r(t + \Delta t) = r(t) + \Delta t \mathcal{G}(r(t + \Delta t), t + \Delta t), \tag{3.5}$$

which yields an *implicit* expression, which can be nonlinear in $r(t + \Delta t)$, depending on $\mathcal{G}$. This is often referred to as a *backward Euler* scheme. These two techniques illustrate the most basic time-stepping schemes used in the scientific community, which form the foundation for the majority of more sophisticated methods. Two main observations can be made:

- The implicit method usually requires one to solve a (nonlinear) equation in $r(t + \Delta t)$.
- The explicit method has the major drawback that the step size Δt may have to be very small to achieve acceptable numerical results. Therefore, an explicit simulation will usually require many more time steps than an implicit simulation.

3.2 An example of stability limitations

Generally speaking, a key difference between the explicit and implicit schemes is their stability properties. By *stability*, we mean that errors made at one stage of the calculations do not cause increasingly larger errors as the computations are continued. For illustration purposes, consider applying each method to the linear scalar differential equation

$$\dot{r} = -cr, \tag{3.6}$$

where $r(0) = r_o$ and c is a positive constant. The exact solution is $r(t) = r_o e^{-ct}$. For the explicit method,

$$\dot{r} \approx \frac{r(t+\Delta t) - r(t)}{\Delta t} = -cr(t), \tag{3.7}$$

which leads to the time-stepping scheme

$$r(L\Delta t) = r_o(1 - c\Delta t)^L, \tag{3.8}$$

where L indicates the time step counter, $t = L\Delta t$ for uniform time steps (as in this example), and $\boldsymbol{r}^L \stackrel{\text{def}}{=} \boldsymbol{r}(t)$, etc. It is stable if $|1 - c\Delta t| < 1$. For the implicit method,

$$\dot{r} \approx \frac{r(t+\Delta t) - r(t)}{\Delta t} = -cr(t+\Delta t), \tag{3.9}$$

which leads to the time-stepping scheme

$$r(L\Delta t) = \frac{r_o}{(1 + c\Delta t)^L}. \tag{3.10}$$

Since $\frac{1}{1+c\Delta t} < 1$, it is always stable. Note that the approximation in Equation (3.8) oscillates in an artificial, nonphysical manner when

$$\Delta t > \frac{2}{c}. \tag{3.11}$$

If $c \gg 1$, then Equation (3.6) is a so-called stiff equation, and Δt may have to be very small for the explicit method to be stable, while, for this example, a larger value of Δt can be used with the implicit method. This motivates the use of implicit methods, with adaptive time stepping, which will be used throughout the remaining analysis.

3.3 Application to particulate flows

Implicit time-stepping methods, with time step size adaptivity, built on approaches found in Zohdi [209], will be used throughout the upcoming analysis. Accordingly, after time discretization of the acceleration term in the equations of motion for a particle (Equation (3.1)),

$$\ddot{\boldsymbol{r}}_i^{L+1} \approx \frac{\boldsymbol{r}_i^{L+1} - 2\boldsymbol{r}_i^L + \boldsymbol{r}_i^{L-1}}{(\Delta t)^2}, \tag{3.12}$$

one arrives at the following abstract form, for the entire system of particles:

$$\mathcal{A}(\boldsymbol{r}^{L+1}) = \mathcal{F}. \tag{3.13}$$

It is convenient to write

$$\mathcal{A}(\boldsymbol{r}^{L+1}) - \mathcal{F} = \mathcal{G}(\boldsymbol{r}^{L+1}) - \boldsymbol{r}^{L+1} + \mathcal{R} = \boldsymbol{0}, \tag{3.14}$$

where $\mathcal{R}$ is a remainder term that does not depend on the solution, i.e.,

$$\mathcal{R} \neq \mathcal{R}(\boldsymbol{r}^{L+1}). \tag{3.15}$$

A straightforward iterative scheme can be written as

$$\boldsymbol{r}^{L+1,K} = \mathcal{G}(\boldsymbol{r}^{L+1,K-1}) + \mathcal{R}, \tag{3.16}$$

where $K = 1, 2, 3, \ldots$ is the index of iteration within time step $L + 1$. The convergence of such a scheme depends on the behavior of $\mathcal{G}$. Namely, a sufficient condition for convergence is that $\mathcal{G}$ be a contraction mapping for all $\boldsymbol{r}^{L+1,K}$, $K = 1, 2, 3, \ldots$. In order to investigate this further, we define the iteration error as

$$\varpi^{L+1,K} \stackrel{\text{def}}{=} \boldsymbol{r}^{L+1,K} - \boldsymbol{r}^{L+1}. \tag{3.17}$$

A necessary restriction for convergence is iterative self-consistency, i.e., the "exact" (discretized) solution must be represented by the scheme

$$\mathcal{G}(\boldsymbol{r}^{L+1}) + \mathcal{R} = \boldsymbol{r}^{L+1}. \tag{3.18}$$

Enforcing this restriction, a sufficient condition for convergence is the existence of a contraction mapping

$$\begin{aligned} ||\varpi^{L+1,K}|| &= ||\boldsymbol{r}^{L+1,K} - \boldsymbol{r}^{L+1}|| \\ &= ||\mathcal{G}(\boldsymbol{r}^{L+1,K-1}) - \mathcal{G}(\boldsymbol{r}^{L+1})|| \leq \eta^{L+1,K} ||\boldsymbol{r}^{L+1,K-1} - \boldsymbol{r}^{L+1}||, \end{aligned} \tag{3.19}$$

where, if

$$0 \leq \eta^{L+1,K} < 1 \tag{3.20}$$

for each iteration K, then

$$\varpi^{L+1,K} \to \boldsymbol{0} \tag{3.21}$$

for any arbitrary starting value $\boldsymbol{r}^{L+1,K=0}$, as $K \to \infty$. This type of contraction condition is sufficient, but not necessary, for convergence. In order to control convergence, we modify the discretization of the acceleration term:[17]

$$\ddot{\boldsymbol{r}}^{L+1} \approx \frac{\dot{\boldsymbol{r}}^{L+1} - \dot{\boldsymbol{r}}^{L}}{\Delta t} \approx \frac{\frac{\boldsymbol{r}^{L+1} - \boldsymbol{r}^{L}}{\Delta t} - \dot{\boldsymbol{r}}^{L}}{\Delta t} \approx \frac{\boldsymbol{r}^{L+1} - \boldsymbol{r}^{L}}{\Delta t^2} - \frac{\dot{\boldsymbol{r}}^{L}}{\Delta t}. \tag{3.22}$$

Inserting this into

$$m\ddot{\boldsymbol{r}} = \boldsymbol{\Psi}^{tot}(\boldsymbol{r}) \tag{3.23}$$

[17]This collapses to a stencil of $\ddot{\boldsymbol{r}}^{L+1} = \frac{\boldsymbol{r}^{L+1} - 2\boldsymbol{r}^{L} + \boldsymbol{r}^{L-1}}{(\Delta t)^2}$ when the time step size is uniform.

leads to

$$\boldsymbol{r}^{L+1,K} \approx \underbrace{\frac{\Delta t^2}{m}\left(\boldsymbol{\Psi}^{tot}(\boldsymbol{r}^{L+1,K-1})\right)}_{\mathcal{G}(\boldsymbol{r}^{L+1,K-1})} + \underbrace{\left(\boldsymbol{r}^{L} + \Delta t \dot{\boldsymbol{r}}^{L}\right)}_{\mathcal{R}}, \tag{3.24}$$

whose convergence is restricted by

$$\eta \propto EIG(\mathcal{G}) \propto \frac{\Delta t^2}{m}. \tag{3.25}$$

Therefore, we see that the eigenvalues of $\mathcal{G}$ are (1) directly dependent on the strength of the interaction forces, (2) inversely proportional to the mass, and (3) directly proportional to $(\Delta t)^2$ (at time $= t$). Therefore, if convergence is slow within a time step, the time step size, which is adjustable, can be reduced by an appropriate amount to increase the rate of convergence. Thus, decreasing the time step size improves the convergence; however, we want to simultaneously maximize the time step sizes to decrease overall computing time while still meeting an error tolerance on the numerical solution's accuracy. In order to achieve this goal, we follow an approach found in Zohdi [208], [209], originally developed for continuum thermochemical multifield problems in which (1) one approximates

$$\eta^{L+1,K} \approx S(\Delta t)^p \tag{3.26}$$

(S is a constant) and (2) one assumes that the error within an iteration behaves according to

$$(S(\Delta t)^p)^K ||\varpi^{L+1,0}|| = ||\varpi^{L+1,K}||, \tag{3.27}$$

$K = 1, 2, \ldots$, where $||\varpi^{L+1,0}||$ is the initial norm of the iterative error and S is intrinsic to the system.[18] Our goal is to meet an error tolerance in exactly a preset number of iterations. To this end, we write

$$(S(\Delta t_{\text{tol}})^p)^{K_d} ||\varpi^{L+1,0}|| = TOL, \tag{3.28}$$

where TOL is a tolerance and K_d is the number of desired iterations.[19] If the error tolerance is not met in the desired number of iterations, the contraction constant $\eta^{L+1,K}$ is too large. Accordingly, one can solve for a new smaller step size under the assumption that S is constant:

$$\Delta t_{\text{tol}} = \Delta t \left(\frac{\left(\frac{TOL}{||\varpi^{L+1,0}||}\right)^{\frac{1}{pK_d}}}{\left(\frac{||\varpi^{L+1,K}||}{||\varpi^{L+1,0}||}\right)^{\frac{1}{pK}}} \right). \tag{3.29}$$

The assumption that S is constant is not critical, since the time steps are to be recursively refined and unrefined throughout the simulation. Clearly, the expression in Equation (3.29) can also be used for time step enlargement if convergence is met in fewer than K_d iterations.

Remark. Time step size adaptivity is important, since the flow's dynamics can dramatically change over the course of time, possibly requiring quite different time step sizes to control the iterative error. However, to maintain the accuracy of the time-stepping scheme, one must respect an upper bound dictated by the discretization error, i.e., $\Delta t \leq \Delta t^{lim}$.

[18] For the class of problems under consideration, due to the quadratic dependency on Δt, typically $p \approx 2$.
[19] Typically, K_d is chosen to be between five and ten iterations.

Remark. Classical solution methods require $\mathcal{O}(N^3)$ operations, whereas iterative schemes, such as the one presented, typically require order N^q, where $1 \leq q \leq 2$. For details, see Axelsson [11]. Also, such solvers are highly advantageous, since solutions to previous time steps can be used as the first guess to accelerate the solution procedure.

Remark. A recursive iterative scheme of Jacobi type, where the updates are made only after one complete system iteration, was illustrated here only for algebraic simplicity. The Jacobi method is easier to address theoretically, while the Gauss–Seidel method, which involves immediately using the most current values, when they become available, is usually used at the implementation level. As is well known, under relatively general conditions, if the Jacobi method converges, the Gauss–Seidel method converges at a faster rate, while if the Jacobi method diverges, the Gauss–Seidel method diverges at a faster rate (for example, see Ames [5] or Axelsson [11]). The iterative approach presented can also be considered as a type of staggering scheme. Staggering schemes have a long history in the computational mechanics community. For example, see Park and Felippa [161], Zienkiewicz [206], Schrefler [173], Lewis et al. [133], Doltsinis [52], [53], Piperno [162], Lewis and Schrefler [132], Armero and Simo [7]–[9], Armero [10], Le Tallec and Mouro [131], Zohdi [208], [209], and the extensive works of Farhat and coworkers (Piperno et al. [163], Farhat et al. [65], Lesoinne and Farhat [130], Farhat and Lesoinne [66], Piperno and Farhat [164], and Farhat et al. [67]).

Remark. It is important to realize that the Jacobi method is perfectly parallelizable. In other words, the calculations for each particle are uncoupled, with the updates only coming afterward. Gauss–Seidel, since it requires the most current updates, couples the particle calculations immediately. However, these methods can be combined to create hybrid approaches whereby the entire particulate flow is partitioned into groups and within each group a Gauss–Seidel method is applied. In other words, for a group, the positions of any particles from outside are initially frozen, as far as calculations involving members of the group are concerned. After each isolated group's solution (particle positions) has converged, computed in parallel, then all positions are updated, i.e., the most current positions become available to all members of the flow, and the isolated group calculations are repeated. See Pöschel and Schwager [167] for a variety of other high-performance techniques, in particular fast contact searches.

Remark. We observe that for the entire ensemble of members one has

$$\sum_{i=1}^{N_p} m_i \ddot{\boldsymbol{r}}_i = \sum_{i=1}^{N_p} \boldsymbol{\Psi}_i^{tot}(\boldsymbol{r}). \tag{3.30}$$

We may decompose the total force due to external sources and internal interaction,

$$\boldsymbol{\Psi}_i^{tot}(\boldsymbol{r}) = \boldsymbol{\Psi}_i^{EXT}(\boldsymbol{r}) + \boldsymbol{\Psi}_i^{INT}(\boldsymbol{r}), \tag{3.31}$$

to obtain

$$\sum_{i=1}^{N_p} m_i \ddot{\boldsymbol{r}}_i = \sum_{i=1}^{N_p} (\boldsymbol{\Psi}_i^{EXT}(\boldsymbol{r}) + \boldsymbol{\Psi}_i^{INT}(\boldsymbol{r})) = \sum_{i=1}^{N_p} \boldsymbol{\Psi}_i^{EXT}(\boldsymbol{r}) + \underbrace{\sum_{i=1}^{N_p} \boldsymbol{\Psi}_i^{INT}(\boldsymbol{r})}_{=\boldsymbol{0}}. \tag{3.32}$$

Thus, a consistency check can be made by tracking the condition

$$\left\| \sum_{i=1}^{N_p} \boldsymbol{\Psi}_i^{INT}(\boldsymbol{r}) \right\| = 0. \tag{3.33}$$

This condition is usually satisfied, to an extremely high level of accuracy, by the previously presented temporally adaptive scheme. However, clearly, this is only a necessary, but not sufficient, condition for zero error.

Remark. An alternative solution scheme would be to attempt to compute the solution by applying a gradient-based method like Newton's method. However, for the class of systems under consideration, there are difficulties with such an approach.

To see this, consider the residual defined by

$$\boldsymbol{\Pi} \stackrel{\text{def}}{=} \mathcal{A}(\boldsymbol{r}) - \mathcal{F}. \tag{3.34}$$

Linearization leads to

$$\boldsymbol{\Pi}(\boldsymbol{r}^K) = \boldsymbol{\Pi}(\boldsymbol{r}^{K-1}) + \nabla_{\boldsymbol{r}}\boldsymbol{\Pi}|_{\boldsymbol{r}^{K-1}}(\boldsymbol{r}^K - \boldsymbol{r}^{K-1}) + \mathcal{O}(||\Delta \boldsymbol{r}||^2), \tag{3.35}$$

and thus the Newton updating scheme can be developed by enforcing

$$\boldsymbol{\Pi}(\boldsymbol{r}^K) \approx \boldsymbol{0}, \tag{3.36}$$

leading to

$$\boldsymbol{r}^K = \boldsymbol{r}^{K-1} - (\mathcal{A}^{TAN,K-1})^{-1}\boldsymbol{\Pi}(\boldsymbol{r}^{K-1}), \tag{3.37}$$

where

$$\mathcal{A}^{TAN,K} = (\nabla_{\boldsymbol{r}}\mathcal{A}(\boldsymbol{r}))\,|_{\boldsymbol{r}^K} = (\nabla_{\boldsymbol{r}}\boldsymbol{\Pi}(\boldsymbol{r}))\,|_{\boldsymbol{r}^K} \tag{3.38}$$

is the tangent. Therefore, in the fixed-point form, one has the operator

$$\mathcal{G}(\boldsymbol{r}) = \boldsymbol{r} - (\mathcal{A}^{TAN})^{-1}\boldsymbol{\Pi}(\boldsymbol{r}). \tag{3.39}$$

For the problems considered, involving contact, friction, near-field forces, etc., it is unlikely that the gradients of $\mathcal{A}$ remain positive definite, or even that $\mathcal{A}$ is continuously differentiable, due to the impact events. Essentially, $\mathcal{A}$ will have nonconvex and nondifferentiable dependence on the positions of the particles. Thus, a fundamental difficulty is the possibility of a zero or nonexistent tangent ($\mathcal{A}^{TAN}$). Therefore, while Newton's method usually converges at a faster rate than a direct fixed-point iteration, quadratically as opposed to superlinearly, its range of applicability is less robust.

3.4 Algorithmic implementation

An implementation of the procedure is given in Algorithm 3.1. The overall goal is to deliver solutions where the iterative error is controlled and the temporal discretization accuracy dictates the upper limit on the time step size (Δt^{lim}).

(1) GLOBAL FIXED-POINT ITERATION (SET $i = 1$ AND $K = 0$):

(2) IF $i > N_p$, THEN GO TO (4);

(3) IF $i \leq N_p$, THEN

(a) COMPUTE POSITION: $\boldsymbol{r}_i^{L+1,K} \approx \frac{\Delta t^2}{m_i}\left(\boldsymbol{\Psi}_i^{tot}(\boldsymbol{r}^{L+1,K-1})\right) + \boldsymbol{r}_i^L + \Delta t \dot{\boldsymbol{r}}_i^L$;

(b) GO TO (2) AND NEXT FLOW PARTICLE ($i = i + 1$);

(4) ERROR MEASURE:

(a) $\varpi_K \stackrel{\text{def}}{=} \frac{\sum_{i=1}^{N_p} ||\boldsymbol{r}_i^{L+1,K} - \boldsymbol{r}_i^{L+1,K-1}||}{\sum_{i=1}^{N_p} ||\boldsymbol{r}_i^{L+1,K} - \boldsymbol{r}_i^L||}$ (normalized);

(b) $Z_K \stackrel{\text{def}}{=} \frac{\varpi_K}{TOL_r}$;

(c) $\Phi_K \stackrel{\text{def}}{=} \left(\frac{(\frac{TOL}{\varpi_0})^{\frac{1}{pK_d}}}{(\frac{\varpi_K}{\varpi_0})^{\frac{1}{pK}}}\right)$;

(5) <u>IF TOLERANCE MET ($Z_K \leq 1$) AND $K < K_d$, THEN</u>

(a) INCREMENT TIME: $t = t + \Delta t$;

(b) CONSTRUCT NEW TIME STEP: $\Delta t = \Phi_K \Delta t$;

(c) SELECT MINIMUM, $\Delta t = \min(\Delta t^{lim}, \Delta t)$, AND GO TO (1);

(6) <u>IF TOLERANCE NOT MET ($Z_K > 1$) AND $K = K_d$, THEN</u>

(a) CONSTRUCT NEW TIME STEP: $\Delta t = \Phi_K \Delta t$;

(b) RESTART AT TIME $= t$ AND GO TO (1).

ALGORITHM 3.1

Remark. At the implementation level in Algorithm 3.1, normalized (nondimensional) error measures were used. As with the unnormalized case, one approximates the error within an iteration to behave according to

$$(S(\Delta t)^p)^K \underbrace{\frac{||\boldsymbol{r}^{L+1,1} - \boldsymbol{r}^{L+1,0}||}{||\boldsymbol{r}^{L+1,0} - \boldsymbol{r}^L||}}_{\varpi_0} = \underbrace{\frac{||\boldsymbol{r}^{L+1,K} - \boldsymbol{r}^{L+1,K-1}||}{||\boldsymbol{r}^{L+1,K} - \boldsymbol{r}^L||}}_{\varpi_K}, \tag{3.40}$$

$K = 2, \ldots,$ where the normalized measures characterize the ratio of the iterative error within a time step to the difference in solutions between time steps. Since both $||\boldsymbol{r}^{L+1,0} - \boldsymbol{r}^L|| \approx \mathcal{O}(\Delta t)$ and $||\boldsymbol{r}^{L+1,K} - \boldsymbol{r}^L|| \approx \mathcal{O}(\Delta t)$ are of the same order, the use of normalized or unnormalized measures makes little difference in rates of convergence. However, the normalized measures are preferred since they have a clearer interpretation.

Remark. Convergence of an iterative scheme can sometimes be accelerated by *relaxation* methods. The basic idea in relaxation methods is to introduce a relaxation parameter,

γ, into the iterations:

$$\boldsymbol{r}^{L+1,K} = \gamma(\mathcal{G}(\boldsymbol{r}^{L+1,K-1}) + \mathcal{R}) + (1-\gamma)\boldsymbol{r}^{L+1,K-1}. \tag{3.41}$$

Since the scheme must reproduce the exact solution, we have

$$\boldsymbol{r}^{L+1} = \gamma(\mathcal{G}(\boldsymbol{r}^{L+1}) + \mathcal{R}) + (1-\gamma)\boldsymbol{r}^{L+1}. \tag{3.42}$$

Subtracting Equation (3.42) from Equation (3.41) yields

$$\boldsymbol{r}^{L+1,K} - \boldsymbol{r}^{L+1} = \gamma\left(\mathcal{G}(\boldsymbol{r}^{L+1,K-1}) - \mathcal{G}(\boldsymbol{r}^{L+1})\right) + (1-\gamma)(\boldsymbol{r}^{L+1,K-1} - \boldsymbol{r}^{L+1}). \tag{3.43}$$

One then forms

$$||\boldsymbol{r}^{L+1,K} - \boldsymbol{r}^{L+1}|| \leq \eta^{\gamma}||\boldsymbol{r}^{L+1,K-1} - \boldsymbol{r}^{L+1}||, \tag{3.44}$$

where the parameter γ is chosen such that $\eta^{\gamma} \leq \eta$, i.e., to induce faster convergence, relative to a relaxation-free approach. The primary difficulty is that the selection of which γ to induce faster convergence is unknown a priori. For even the linear theory, i.e., when $\mathcal{G}$ is a linear operator, such parameters are unknown and are usually computed by empirical trial and error procedures. See Axelsson [11] for reviews.

Remark. There are alternative ways of accelerating convergence. As we recall, geometric convergence of the sequence $a^1, a^2, \ldots, a^K, \ldots, a$ implies

$$\frac{a - a^{K+1}}{a - a^K} = \varrho < 1, \tag{3.45}$$

where ϱ is a constant and a is the limit. Now consider the following sequence of terms:

$$\boxed{\begin{aligned} &a \approx a^K + C\varrho^K \Rightarrow a - a^K \approx C\varrho^K, \\ &\Rightarrow a - a^{K+1} \approx C\varrho^{K+1} = \varrho(a - a^K), \\ &\Rightarrow a - a^{K+2} \approx C\varrho^{K+2} = \varrho(a - a^{K+1}), \end{aligned}} \tag{3.46}$$

where C is a constant. These equations can be solved simultaneously to yield

$$a \approx \frac{a^{K+2}a^K - (a^K)^2}{a^{K+2} + a^K - 2a^{K+1}}. \tag{3.47}$$

If Equation (3.45) were true, then the value of a computed from Equation (3.47) would be exact for all K. Only in rare cases will it be true, so we construct a new sequence, for all K, from the old one:

$$a^{K,1} = \frac{a^{K+2}a^K - (a^K)^2}{a^{K+2} + a^K - 2a^{K+1}}. \tag{3.48}$$

We then repeat the procedure on the newly generated sequence:

$$a^{K,2} = \frac{a^{K+2,1}a^{K,1} - (a^{K,1})^2}{a^{K+2,1} + a^{K,1} - 2a_i^{K+1,1}}. \tag{3.49}$$

With each successive extrapolation, the new sequence becomes two members shorter than the previous one. We repeat the procedure until the sequence is only one member long. The final member is an approximation to the limit. It is remarked that the initial sequence does not even have to be monotone for the process to converge to the true limit. This process is frequently referred to as an Aitken-type extrapolation. For an in-depth analysis of this procedure, see Aitken [4], Shanks [176], or Arfken [6]. Such methods are sometimes useful for extrapolating smooth numerical solutions to differential equations.

Chapter 4

Representative numerical simulations

In order to illustrate how to simulate a particulate flow, we consider a group of N_p randomly positioned particles in a cubical domain with dimensions $D \times D \times D$. During the simulation, if a particle escapes from the control volume, the position component is reversed and the velocity component is retained (now incoming). Thus, for example, if the x_1 component of the position vector for the ith particle exceeds the boundary of the control volume, then $r_{ix_1} = -r_{ix_1}$ is enforced. These boundary conditions are sometimes referred to as "periodic" boundary conditions.[20] The particle size and volume fraction occupied are determined by a particle/sample size ratio, which is defined via a "subvolume" size[21]

$$V \stackrel{\text{def}}{=} \frac{D \times D \times D}{N_p}. \tag{4.1}$$

The ratio between the particle radii (assumed the same for this example), denoted by b, and the subvolume is

$$\mathcal{L} \stackrel{\text{def}}{=} \frac{b}{V^{\frac{1}{3}}}. \tag{4.2}$$

The volume fraction occupied by the particles is

$$v_f \stackrel{\text{def}}{=} \frac{4\pi \mathcal{L}^3}{3}. \tag{4.3}$$

Thus, the total volume occupied by the particles, denoted by Δ, can be written as

$$\nu = v_f N_p V, \tag{4.4}$$

and the total mass is

$$M = \sum_{i=1}^{N_p} m_i = \rho \nu, \tag{4.5}$$

[20]There are many variants of this procedure.

[21] D is normalized to unity in these simulations.

while that of an individual particle, assuming that all are the same size, is

$$m_i = \frac{\rho v}{N_p} = \rho \frac{4}{3} \pi b_i^3. \tag{4.6}$$

Remark. In the upcoming simulations, the classical random sequential addition algorithm was used to place nonoverlapping particles into the computational domain (Widom [200]). This algorithm was adequate for the volume fraction ranges of interest (under 30%), since its limit is on the order of 38%. To achieve higher volume fractions, there are several more sophisticated algorithms, such as the classical equilibrium-based Metropolis algorithm. For a detailed review of a variety of such methods, see Torquato [194]. For much higher volume fractions, effectively packing (and "jamming") particles to theoretical limits (approximately 74%), a new class of methods has recently been developed, based on simultaneous particle flow and growth, by Torquato and coworkers (see, for example, Kansaal et al. [119] and Donev et al. [55]–[59]). This class of methods was not employed in the present study due to the relatively moderate volume fraction range of interest here; however, such methods appear to offer distinct computational advantages if extremely high volume fractions are desired.

4.1 Simulation parameters

The relevant simulation parameters were

- number of particles $= 100$,
- (normalized) box dimension $D = 1$ m,
- initial mean velocity field $= (1.0, 0.1, 0.1)$ m/s,
- initial random perturbations around mean velocity $= (\pm 1.0, \pm 0.1, \pm 0.1)$ m/s,
- (normalized) length scale of the particles, $\mathcal{L} = 0.25$, with corresponding volume fraction $v_f = \frac{4\pi\mathcal{L}^3}{3} = 0.0655$ and radius $b = 0.0539$ m,
- mass density of the particles $= 2000$ kg/m^3,
- simulation duration $= 1$ s,
- initial time step size $= 0.001$ s,
- time step upper bound $= 0.01$ s,
- tolerance for the fixed-point iteration $= 10^{-6}$.

The parameters $\overline{\alpha}_1$ and $\overline{\alpha}_2$, which represent the strength of the near-field interaction forces per unit mass[2], were varied to investigate the near-field effects on the overall particulate flow. During the simulations, we enforced the stability condition in Equation (1.37) by setting $(\beta_1, \beta_2) = (1, 2)$.

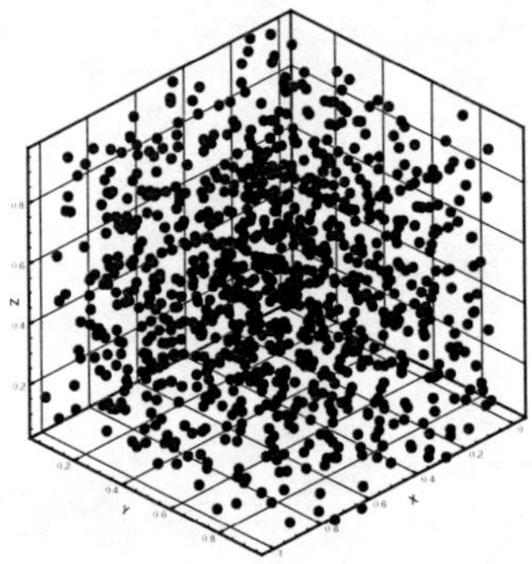

Figure 4.1. *A typical starting configuration for the types of particulate systems under consideration.*

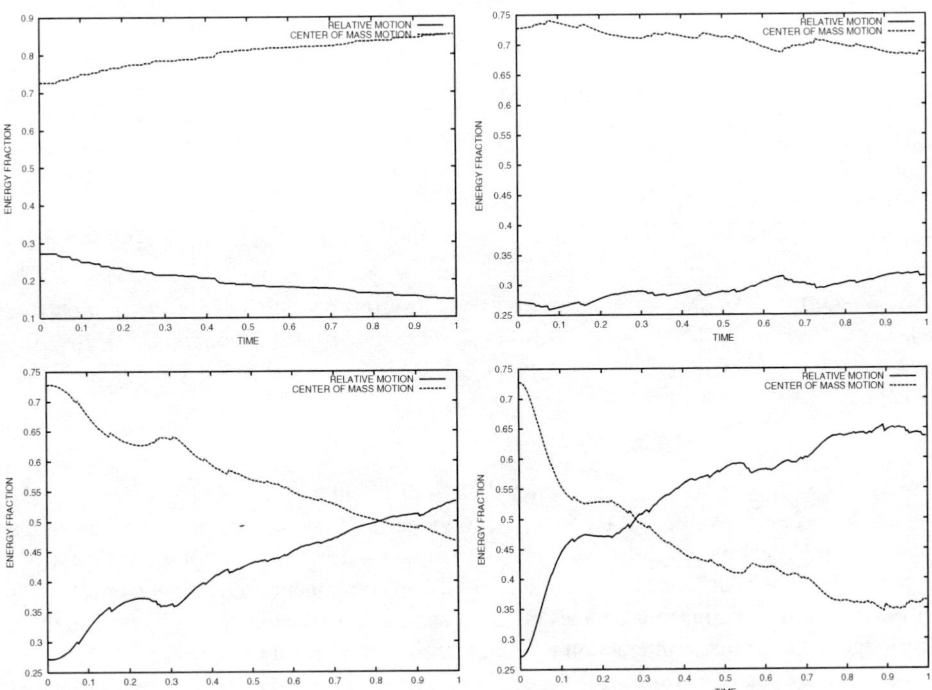

Figure 4.2. *The proportions of the kinetic energy that are bulk and relative motion. Top to bottom and left to right, for* $e_o = 0.5$, $\mu_s = 0.2$, $\mu_d = 0.1$*:* (1) *no near-field interaction,* (2) $\overline{\alpha}_1 = 0.1$ *and* $\overline{\alpha}_2 = 0.05$, (3) $\overline{\alpha}_1 = 0.25$ *and* $\overline{\alpha}_2 = 0.125$, *and* (4) $\overline{\alpha}_1 = 0.5$ *and* $\overline{\alpha}_2 = 0.25$ *(Zohdi* [212]*).*

4.2 Results and observations

The starting configuration is shown in Figure 4.1. Figures 4.2 and 4.3 illustrate the computational results. The type of motion, characterized by the proportions of bulk and rela-

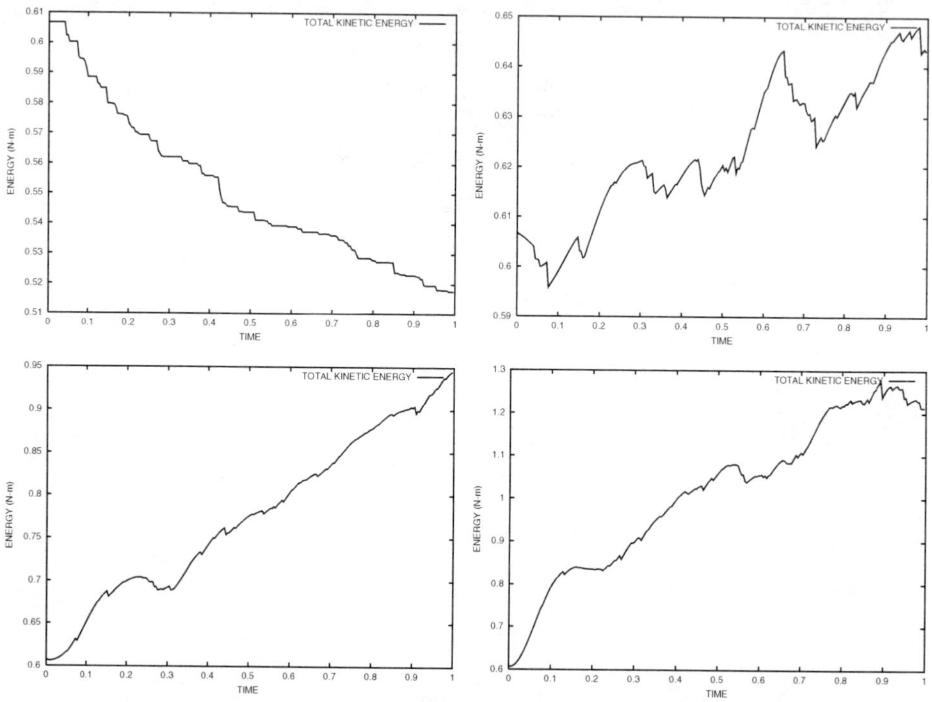

Figure 4.3. *The total kinetic energy in the system per unit mass. Top to bottom and left to right, for* $e_o = 0.5$, $\mu_s = 0.2$, $\mu_d = 0.1$*:* (1) *no near-field interaction,* (2) $\overline{\alpha}_1 = 0.1$ *and* $\overline{\alpha}_2 = 0.05$, (3) $\overline{\alpha}_1 = 0.25$ *and* $\overline{\alpha}_2 = 0.125$, *and* (4) $\overline{\alpha}_1 = 0.5$ *and* $\overline{\alpha}_2 = 0.25$ *(Zohdi* [212]*).*

tive kinetic energy in the system, is dramatically different with increasing severity of the near-field forces.[22] Notice that the kinetic energy per unit mass is nonmonotone when the near-field interactions are taken into account (Figure 4.3). One may observe that, from Figure 4.2, as the near-field strength is increased, the component of the kinetic energy corresponding to the relative motion does not decay and actually becomes dominant with time. Essentially, the near-field interaction becomes strong enough that the flowing system experiences a transition to a *vibrating ensemble*. This transition can be qualitatively examined by recognizing that the governing equations are formally similar to classical, normalized, linear (or linearized) second-order equations governing a one degree of freedom harmonic oscillator of the form

$$\ddot{r} + 2\zeta\omega_n\dot{r} + \omega_n^2 r = \frac{f(t)}{m}, \tag{4.7}$$

where

$$\omega_n = \sqrt{\frac{k}{m}}, \tag{4.8}$$

[22] Typically, the simulations took under a minute on a single laptop.

r is the position measured from equilibrium ($r = 0$), k is the stiffness associated with the restoring force (kr), m represents the mass, and the damping ratio is

$$\zeta \stackrel{\text{def}}{=} \frac{d}{2m\omega_n}, \tag{4.9}$$

d being a constant of damping and $f(t)$ an external forcing term. The damped period of natural, *force-free* vibration is

$$\mathcal{T}_d \stackrel{\text{def}}{=} \frac{2\pi}{\omega_d}, \tag{4.10}$$

where

$$\omega_d \stackrel{\text{def}}{=} \omega_n\sqrt{1-\zeta^2} \tag{4.11}$$

is the "damped natural frequency." Using standard procedures, one decomposes the solution into homogeneous and particular parts:

$$r = r_H + r_P. \tag{4.12}$$

The homogeneous part must satisfy

$$\ddot{r}_H + 2\zeta\omega_n\dot{r}_H + \omega_n^2 r_H = 0. \tag{4.13}$$

Assuming the standard form

$$r_H = \exp(\lambda t) \tag{4.14}$$

yields, upon substitution,

$$\lambda^2 \exp(\lambda t) + 2\zeta\omega_n\lambda\exp(\lambda t) + \omega_n^2\exp(\lambda t) = 0, \tag{4.15}$$

leading to the characteristic equation

$$\lambda^2 + 2\zeta\omega_n\lambda + \omega_n^2 = 0. \tag{4.16}$$

Solving for the roots yields

$$\lambda_{1,2} = \omega_n(-\zeta \pm \sqrt{\zeta^2 - 1}). \tag{4.17}$$

The general solution is

$$r = A_1\exp(\lambda_1 t) + A_2\exp(\lambda_2 t). \tag{4.18}$$

Depending on the value of ζ, the solution will have one of three distinct types of behavior:

- $\zeta > 1$, overdamped, leading to no oscillation, where the value of r approaches zero for large values of time. Mathematically, λ_1 and λ_2 are negative numbers, so

$$r_H = A_1\exp(\omega_n(-\zeta + \sqrt{\zeta^2-1})t) + A_2\exp(\omega_n(-\zeta - \sqrt{\zeta^2-1})t). \tag{4.19}$$

- $\zeta = 1$, critically damped, leading to no oscillation, where the value of r approaches zero for large values of time, but faster than the overdamped solution. Mathematically, λ_1 and λ_2 are equal real numbers, $\lambda_1 = \lambda_2 = -\omega_n$, so

$$r_H = (A_1 + A_2 t)\exp(\omega_n t). \tag{4.20}$$

- $\zeta < 1$, underdamped, leading to damped oscillation, where the value of r approaches zero for large values of time, in an oscillatory fashion. Mathematically, $\zeta^2 - 1 < 0$, so

$$r_H = A_1 \cos(\omega_d t) + A_2 \sin(\omega_d t). \tag{4.21}$$

Thus, under certain conditions, a particulate flow can vibrate or "pulse." The particular solution, generated by the presence of externally applied forces, satisfies the differential equation for a specific right-hand side:

$$\ddot{r}_P + 2\zeta\omega_n\dot{r}_P + \omega_n^2 r_P = \frac{f(t)}{m}. \tag{4.22}$$

For example, if

$$f(t) = f_o \sin(\Omega t), \tag{4.23}$$

then

$$r_P = R\sin(\Omega t - \phi), \tag{4.24}$$

where

$$R = \frac{f_o}{k\sqrt{\left(1 - \frac{\Omega^2}{\omega_n^2}\right)^2 + \left(2\zeta\frac{\Omega}{\omega_n}\right)^2}} \tag{4.25}$$

and

$$\phi = \tan^{-1}\left(\frac{2\zeta\frac{\Omega}{\omega_n}}{1 - \frac{\Omega^2}{\omega_n^2}}\right). \tag{4.26}$$

Thus, clearly, such systems may resonate if forced at certain frequencies. In order to qualitatively tie this directly to the form of problem considered in this work, consider a linearization of a single nonlinear differential equation, describing the attraction and repulsion from the origin ($r_o = 0$) of the form[23]

$$m\ddot{r} + d\dot{r} = \Psi^{nf}(r), \tag{4.27}$$

where

$$\Psi^{nf}(r) = -\alpha_1 r^{-\beta_1} + \alpha_2 r^{-\beta_2} \tag{4.28}$$

and d is an effective dissipation term that would arise from inelastic impact and friction. Upon linearization of the nonlinear interaction relation about a point r_*,

$$\Psi^{nf}(r) \approx \Psi^{nf}(r_*) + \left.\frac{\partial\Psi^{nf}}{\partial r}\right|_{r=r_*}(r - r_*) + \mathcal{O}(r - r_*), \tag{4.29}$$

and normalizing the equations, we obtain

$$\ddot{r} + 2\zeta^*\omega_n^*\dot{r} + (\omega_n^*)^2 r = \frac{f^*(t)}{m}, \tag{4.30}$$

[23]The unit normal has been taken into account, thus the presence of a change in sign.

where

$$\omega_n^* = \sqrt{\frac{-\frac{\partial \Psi^{nf}}{\partial r}|_{r=r_*}}{m}}, \tag{4.31}$$

$$\zeta^* = \frac{d}{2m\omega_n^*}, \tag{4.32}$$

and

$$f^*(t) = \Psi^{nf}(r_*) - \left.\frac{\partial \Psi^{nf}}{\partial r}\right|_{r=r_*} r_*. \tag{4.33}$$

For the specific interaction form chosen, we have

$$\omega_n^* = \sqrt{\frac{-\alpha_1\beta_1 r_*^{-\beta_1-1} + \alpha_2\beta_2 r_*^{-\beta_2-1}}{m}} = \sqrt{-\overline{\alpha}_1 m\beta_1 r_*^{-\beta_1-1} + \overline{\alpha}_2 m\beta_2 r_*^{-\beta_2-1}}, \tag{4.34}$$

where the "loading" is

$$f^*(t) = -\alpha_1 r_*^{-\beta_1} + \alpha_2 r_*^{-\beta_2} - \alpha_1\beta_1 r_*^{-\beta_1-1} + \alpha_2\beta_2 r_*^{-\beta_2-1}. \tag{4.35}$$

We note that if the parameters are chosen (as in the preceding simulations) specifically as $(\beta_1, \beta_2) = (1, 2)$ and r_* is chosen as the static equilibrium point, r_e, given by Equation (1.36), then

$$r_* = r_e = \frac{\alpha_2}{\alpha_1} \tag{4.36}$$

and

$$\omega_n^* = \sqrt{\frac{\alpha_1\left(\frac{\alpha_1}{\alpha_2}\right)^2}{m}} = \sqrt{\frac{\alpha_1}{m}\left(\frac{\overline{\alpha}_1}{\overline{\alpha}_2}\right)^2} \stackrel{\text{def}}{=} \sqrt{\frac{k^*}{m}}, \tag{4.37}$$

where

$$k^* \stackrel{\text{def}}{=} \alpha_1\left(\frac{\overline{\alpha}_1}{\overline{\alpha}_2}\right)^2. \tag{4.38}$$

Thus, in the preceding numerical examples, when we kept the ratio $\frac{\overline{\alpha}_1}{\overline{\alpha}_2}$ constant, but increased α_1 (while keeping m constant), we were effectively increasing the "stiffness" in the system and, therefore, the amount of (pre)stored energy available to counteract dissipation. Clearly, under certain conditions, a particulate flow may "pulse" (oscillate) depending on the character of the interaction and the contact parameters. Thus, oscillatory behavior is not unexpected for the multibody system (Figure 4.3). We remark that increasingly smaller ω_n^* indicates that the system tends toward a "regular" (near-field–free) particulate flow. Smaller ω_n^* occurs with heavier particles or smaller attractive forces, and larger values of ζ^* (more damped) occur when increased friction or smaller restitution coefficients are present in the flow. Clearly, key dimensionless parameters, like ζ^*, characterize the oscillatory behavior and the fluctuating motion with respect to mean values within the particulate flow.

Chapter 5

Inverse problems/parameter identification

An important aspect of any model is the identification of parameters that force the system behavior to match a (desired) target response. For example, in the ideal case, one would like to determine the type of near-field interaction that produces certain flow characteristics, via numerical simulations, in order to guide or minimize time-consuming laboratory tests. As a representative of a class of model problems, consider inverse problems, where the parameters in the near-field interaction representation are sought, the α's and β's, that deliver a target particulate flow behavior by minimizing a normalized cost function

$$\Pi = \frac{\int_0^{\mathcal{T}} |A - A^*|\,dt}{\int_0^{\mathcal{T}} |A^*|\,dt}, \tag{5.1}$$

where the total simulation time is $\mathcal{T}$, A is a computationally generated quantity of interest, and A^* is the target response. Typically, for the class of problems considered in this work, formulations (Π) such as in Equation (5.1) depend, in a nonconvex and nondifferentiable manner, on the α's and β's. This is primarily due to the nonlinear character of the near-field interaction, the physics of sudden interparticle impact, and the transient dynamics. Clearly, we must have restrictions (for physical reasons) on the parameters in the near-field interaction:

$$\alpha^-_{1\ or\ 2} \leq \alpha_{1\ or\ 2} \leq \alpha^+_{1\ or\ 2} \tag{5.2}$$

and

$$\beta^-_{1\ or\ 2} \leq \beta_{1\ or\ 2} \leq \beta^+_{1\ or\ 2}, \tag{5.3}$$

where $\alpha^-_{1\ or\ 2}$, $\alpha^+_{1\ or\ 2}$, $\beta^-_{1\ or\ 2}$, and $\beta^+_{1\ or\ 2}$ are the lower and upper limits on the coefficients in the interaction forces.[24] With respect to the minimization of Equation (5.1), classical gradient-based deterministic optimization techniques are not robust, due to difficulties with objective function nonconvexity and nondifferentiability. Classical gradient-based algorithms are likely to converge only toward a local minimum of the objective function unless a sufficiently close initial guess to the global minimum is not provided. Also, it is usually

[24]Additionally, we could also vary the other parameters in the system, such as the friction, particle densities, and drag. However, we shall fix these parameters during the upcoming examples.

extremely difficult to construct an initial guess that lies within the (global) convergence radius of a gradient-based method. These difficulties can be circumvented by using a certain class of simple, yet robust, nonderivative search methods, usually termed "genetic" algorithms, before applying gradient-based schemes. Genetic algorithms are search methods based on the principles of natural selection, employing concepts of species evolution such as reproduction, mutation, and crossover. Implementation typically involves a randomly generated population of fixed-length elemental strings, "genetic information," each of which represents a specific choice of system parameters. The population of individuals undergo "mating sequences" and other biologically inspired events in order to find promising regions of the search space. There are a variety of such methods, which employ concepts of species evolution, such as reproduction, mutation, and crossover. Such methods can be traced back, at least, to the work of John Holland (Holland [94]). For reviews of such methods, see, for example, Goldberg [77], Davis [50], Onwubiko [155], Kennedy and Eberhart [120], Lagaros et al. [129], Papadrakakis et al. [156]–[160], and Goldberg and Deb [78].

5.1 A genetic algorithm

As examples of objective functions that one might minimize, consider the following:

- overall energetic behavior per unit mass (Equation (2.29)):

$$\Pi_T = \frac{\int_0^{\mathcal{T}} |\overline{T} - \overline{T}^*|\, dt}{\int_0^{\mathcal{T}} \overline{T}^*\, dt}, \tag{5.4}$$

  where the total simulation time is $\mathcal{T}$ and where $\overline{T}^*$ is a target energy per unit mass value;

- energy component distribution (Equation (2.29)):

$$\Pi_{Tr} = \frac{\int_0^{\mathcal{T}} |T_r - T_r^*|\, dt}{\int_0^{\mathcal{T}} T_r^*\, dt} \tag{5.5}$$

  for the relative motion part, and

$$\Pi_{Tb} = \frac{\int_0^{\mathcal{T}} |T_b - T_b^*|\, dt}{\int_0^{\mathcal{T}} T_b^*\, dt} \tag{5.6}$$

  for the bulk motion part, where the fraction of kinetic energy due to relative motion is T_r, the fraction of kinetic energy due to bulk motion is T_b, and T_r^* and T_b^* are the target values.

Compactly, one may write

$$\Pi = \frac{w_T \Pi_T + w_{Tr} \Pi_{Tr} + w_{Tb} \Pi_{Tb}}{w_T + w_{Tr} + w_{Tb}}, \tag{5.7}$$

where the w's are weights.

Adopting the approaches found in Zohdi [209]–[216], a genetic algorithm has been developed to treat nonconvex inverse problems involving various aspects of multiparticle mechanics. The central idea is that the system parameters form a genetic string and a survival of the fittest algorithm is applied to a population of such strings. The overall process is as follows: (a) a population (S) of different parameter sets is generated at random within the parameter space, each represented by a (genetic) string of the system (N) parameters; (b) the performance of each parameter set is tested; (c) the parameter sets are ranked from top to bottom according to their performance; (d) the best parameter sets (parents) are mated pairwise, producing two offspring (children), i.e., each best pair exchanges information by taking random convex combinations of the parameter set components of the parents' genetic strings; and (e) the worst-performing genetic strings are eliminated, new replacement parameter sets (genetic strings) are introduced into the remaining population of best-performing genetic strings, and the process (a)–(e) is then repeated. The term "fitness" of a genetic string is used to indicate the value of the objective function. The most fit genetic string is the one with the smallest objective function. The retention of the most fit genetic strings from a previous generation (parents) is critical, since if the objective functions are highly nonconvex (the present case), there exists a clear possibility that the inferior offspring will replace superior parents. When the top parents are retained, the minimization of the cost function is guaranteed to be monotone (guaranteed improvement) with increasing generations. There is no guarantee of successive improvement if the top parents are not retained, even though nonretention of parents allows more new genetic strings to be evaluated in the next generation. In the scientific literature, numerical studies imply that, for sufficiently large populations, the benefits of parent retention outweigh this advantage and any disadvantages of "inbreeding," i.e., a stagnant population (Figure 5.1). For more details on this so-called inheritance property, see Davis [50] or Kennedy and Eberhart [120]. In the upcoming algorithm, inbreeding is mitigated, since, with each new generation, new parameter sets, selected at random within the parameter space, are added to the population. Previous numerical studies by this author (Zohdi [209]–[216]) have indicated that not retaining the parents is suboptimal due to the possibility that inferior offspring will replace

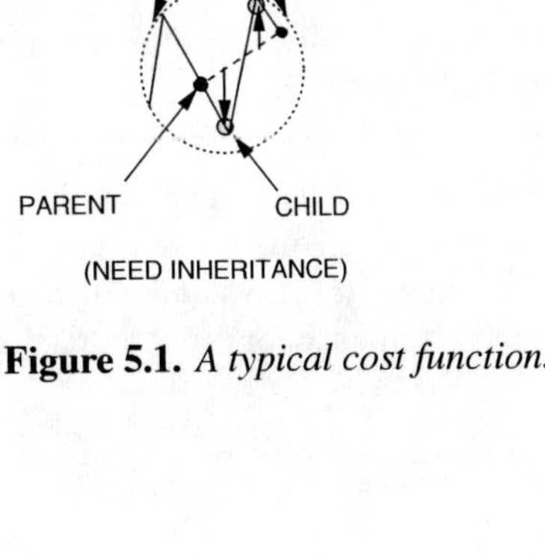

Figure 5.1. *A typical cost function.*

superior parents. Additionally, parent retention is computationally less expensive, since these parameter sets do not have to be reevaluated (or ranked) in the next generation.

An implementation of such ideas is as follows (Zohdi [209]–[216]).

- **STEP 1:** Randomly generate a population of S starting genetic strings, $\mathbf{\Lambda}^i$ ($i = 1, \ldots, S$):

$$\mathbf{\Lambda}^i \stackrel{\text{def}}{=} \{\Lambda_1^i, \Lambda_2^i, \Lambda_3^i, \Lambda_4^i, \ldots, \ldots, \Lambda_N^i\} = \{\overline{\alpha}_1^i, \beta_1^i, \overline{\alpha}_2^i, \beta_2^i, \ldots\}.$$

- **STEP 2:** Compute the fitness of each string $\Pi(\mathbf{\Lambda}^i)$ ($i = 1, \ldots, S$).
- **STEP 3:** Rank genetic strings: $\mathbf{\Lambda}^i$ ($i = 1, \ldots, S$).
- **STEP 4:** Mate the nearest pairs and produce two offspring ($i = 1, \ldots, S$):

$$\boldsymbol{\lambda}^i \stackrel{\text{def}}{=} \Phi^{(I)}\mathbf{\Lambda}^i + (1 - \Phi^{(I)})\mathbf{\Lambda}^{i+1}, \quad \boldsymbol{\lambda}^{i+1} \stackrel{\text{def}}{=} \Phi^{(II)}\mathbf{\Lambda}^i + (1 - \Phi^{(II)})\mathbf{\Lambda}^{i+1}.$$

- **NOTE:** $\Phi^{(I)}$ and $\Phi^{(II)}$ are random numbers, such that $0 \leq \Phi^{(I)}, \Phi^{(II)} \leq 1$, which are different for each component of each genetic string.
- **STEP 5:** Kill off the bottom $M < S$ strings and keep the top $K < N$ parents and the top K offspring (K offspring $+ K$ parents $+ M = S$).
- **STEP 6:** Repeat Steps 1–6 with the top gene pool (K offspring and K parents), plus M new, randomly generated, strings.
- **OPTION:** Rescale and restart the search around the best-performing parameter set every few generations.
- **OPTION:** We remark that gradient-based methods are sometimes useful for postprocessing solutions found with a genetic algorithm if the objective function is sufficiently smooth in that region of the parameter space. In other words, if one has located the convex portion of the parameter space with a global genetic search, one can employ gradient-based procedures locally to minimize the objective function further. In such procedures, in order to obtain a new directional step for $\mathbf{\Lambda}$, one must solve the system

$$[\mathbb{H}]\{\Delta\mathbf{\Lambda}\} = -\{\boldsymbol{g}\}, \tag{5.8}$$

where $[\mathbb{H}]$ is the Hessian matrix ($N \times N$), $\{\Delta\mathbf{\Lambda}\}$ is the parameter increment ($N \times 1$), and $\{\boldsymbol{g}\}$ is the gradient ($N \times 1$). We shall not employ this second (postgenetic) stage in this work. An exhaustive review of these methods can be found in the texts of Luenberger [142] and Gill et al. [76], while the state of the art can be found in Papadrakakis et al. [160].

Remark. It is important to scale the system variables, for example, to be positive numbers and of comparable magnitude, in order to avoid dealing with large variations in the parameter vector components. Typically, for systems with a finite number of particles, there will be slight variations in the performance for different random starting configurations. In order to stabilize the objective function's value with respect to the randomness of the flow starting configuration, for a given parameter selection ($\mathbf{\Lambda}$, characterized by the α's and β's), a regularization procedure is applied within the genetic algorithm, whereby the performances of a series of different random starting configurations are averaged until the (ensemble) average converges, i.e., until the following condition is met:

$$\left|\frac{1}{E+1}\sum_{i=1}^{E+1}\Pi^{(i)}(\mathbf{\Lambda}^I)-\frac{1}{E}\sum_{i=1}^{E}\Pi^{(i)}(\mathbf{\Lambda}^I)\right|\leq TOL\left|\frac{1}{E+1}\sum_{i=1}^{E+1}\Pi^{(i)}(\mathbf{\Lambda}^I)\right|, \qquad (5.9)$$

where index i indicates a different starting random configuration ($i = 1, 2, \ldots, E$) that has been generated and E indicates the total number of configurations tested. In order to implement this in the genetic algorithm, in Step 2, one simply replaces *compute* with *ensemble compute*, which requires a further inner loop to test the performance of multiple starting configurations. Similar ideas have been applied to randomly dispersed particulate media with solid binders in Zohdi [209]–[216].

5.2 A representative example

We considered a search space of $0 \leq \overline{\alpha}_1 \leq 1$, $0 \leq \beta_1 \leq 1$, $0 \leq \overline{\alpha}_2 \leq 1$, and $1 \leq \beta_2 \leq 2$. Recall that the stability restriction on the exponents was $\frac{\beta_2}{\beta_1} > 1$, thus motivating the choice of the range of search. As in the previous simulations, 100 particles with periodic boundary conditions were used. The total time was set to be 1 s ($\mathcal{T} = 1$). The starting state values of the system were the same as in the previous examples. The target objective (behavior) values were constants: $(\overline{T}^*, T_b^*, T_r^*) = (1.0, 0.5, 0.5)$. Such an objective can be interpreted as forcing a system with given initial behavior to adapt to a different type of behavior within a given time interval. The number of genetic strings in the population was set to 20, for 20 generations, allowing 6 total offspring of the top 6 parents (2 from each parental pair), along with their parents, to proceed to the next generation. Therefore, after each generation, 8 entirely new (randomly generated) genetic strings are introduced. Every 10 generations, the search was rescaled around the best parameter set and the search restarted. Figure 5.2 and Table 5.1 depict the results. A total of 310 parameter selections were tested. The total number of strings tested was 1757, thus requiring an average of 5.68 strings per parameter selection for the ensemble-averaging stabilization. The behavior of the best parameter selection's response is shown in Figure 5.3.

Table 5.1. *The optimal coefficients of attraction and repulsion for the particulate flow and the top six fitnesses.*

Rank	$\overline{\alpha}_1$	β_1	$\overline{\alpha}_2$	β_2	Π
1	0.35935	0.67398	0.25659	1.58766	0.065228
2	0.31214	0.67816	0.22113	1.65054	0.065690
3	0.30032	0.54474	0.22240	1.51649	0.070433
4	0.31143	0.57278	0.25503	1.36696	0.073200
5	0.32872	0.74653	0.25560	1.56315	0.078229
6	0.30580	0.74276	0.27228	1.36962	0.090701

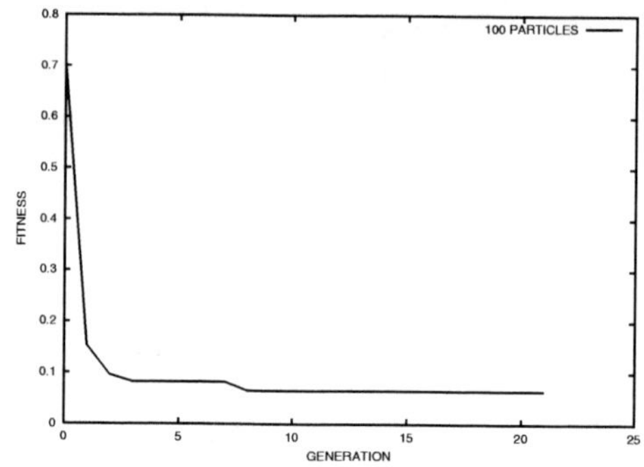

Figure 5.2. *The best parameter set's* ($\overline{\alpha}_1, \overline{\alpha}_2, \beta_1, \beta_2$) *objective function value with passing generations (Zohdi* [212]*).*

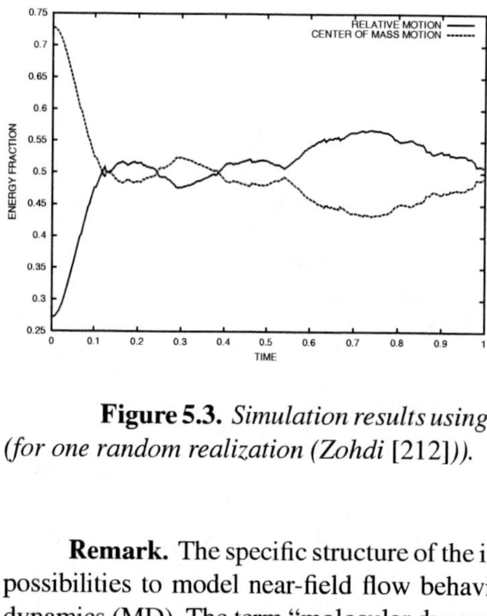

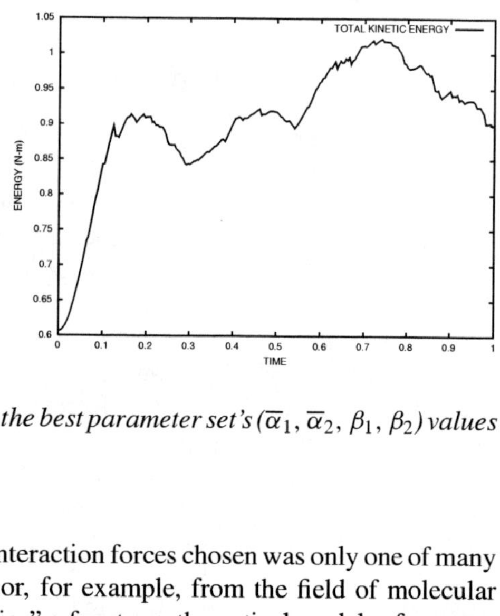

Figure 5.3. *Simulation results using the best parameter set's* ($\overline{\alpha}_1, \overline{\alpha}_2, \beta_1, \beta_2$) *values (for one random realization (Zohdi* [212]*)).*

Remark. The specific structure of the interaction forces chosen was only one of many possibilities to model near-field flow behavior, for example, from the field of molecular dynamics (MD). The term "molecular dynamics" refers to mathematical models of systems of atoms or molecules where each atom (or molecule) is represented by a material point in R^3 and is treated as a point mass. The overall motion of such mass-point systems is dictated by Newtonian mechanics. For an extensive survey of MD-type interaction forces, which includes comparisons of the theoretical and computational properties of each interaction law, we refer the reader to Frenklach and Carmer [71]. MD is typically used to calculate (ensemble) averages of thermochemical and thermomechanical properties of gases, liquids, or solids. The analogy between particulate flow dynamics and MD of an atomistic chemical system is inescapable. In the usual MD approach (see Haile [87], for example), the motion of individual atoms is described by Newton's second law with the forces computed from a prescribed potential energy function, $V(\boldsymbol{r})$, $m\ddot{\boldsymbol{r}} = -\nabla V(\boldsymbol{r})$. The MD approach has been applied to describe all material phases: solids, liquids, and gases, as well as biological systems (Hase [89] and Schlick [171]). For instance, a Fourier transform of the velocity autocorrelation function specifies the "bulk" diffusion coefficient (Rapaport [168]). The mathematical form of more sophisticated potentials to produce interaction forces, $\mathbf{\Psi}^{nf} = -\nabla V$, is rooted in the expansion

$$V = \sum_{i,j} V_2 + \sum_{i,j,k} V_3 + \cdots, \tag{5.10}$$

where V_2 is the binary, V_3 the tertiary, etc., potential energy function, and the summations are taken over corresponding combinations of atoms. The binary functions usually take the form of the familiar Mie, Lennard–Jones, and Morse potentials (Moelwyn-Hughes [149]). The expansions beyond the binary interactions introduce either three-body terms directly (Stillinger and Weber [179]) or as "local" modifications of the two-body terms (Tersoff [193]). Clearly, the inverse parameter identification technique presented is applicable to such representations, but with more adjustable search parameters. For examples with significantly more search parameter complexity, see Zohdi [209]–[216].

Chapter 6

Extensions to "swarm-like" systems

It is important to realize that nontraditional particulate-like models are frequently used to simulate the behavior of groups comprising individual units whose interaction is represented by near-field interaction forces. The basis of such interaction is not a "charge."[25] As an example, we provide an introduction to an emerging field, closely related to dry particulate flows, that has relatively recently received considerable attention, namely, the analysis of *swarms*. In a very general sense, the term "swarm" is usually meant to signify any collection of objects (agents) that interact with one another. It has long been recognized that interactive cooperative behavior within biological groups or swarms is advantageous in avoiding predators or, vice versa, in capturing prey. For example, one of the primary advantages of a swarm-like decentralized decision-making structure is that there is no leader and thus the vulnerability of the swarm is substantially reduced. Furthermore, the decision making is relatively simple and rapid for each individual; however, the aggregate behavior of the swarm can be quite sophisticated. Although the modeling of swarm-like behavior has biological research origins, dating back at least to Breder [36], it can be treated as a purely multiparticle dynamical system, where the communication between swarm members is modeled via interaction forces. It is commonly accepted that a central characteristic of swarm-like behavior is the tradeoff between long-range interaction and short-range repulsion between individuals. Models describing clouds or swarms of particles, where their interaction is constructed from attractive and repulsive forces, dependent on the relative distance between individuals, are commonplace. For reviews, see Gazi and Passino [75], Bender and Fenton [25], or Kennedy and Eberhart [120]. The field is quite large and encompasses a wide variety of applications, for example, the behavior of flocks of birds, schools of fish, flow of traffic, and crowds of human beings, to name a few. Loosely speaking, swarm analyses are concerned with the complex aggregate behavior of groups of simple members, which are frequently treated as particles (for example, in Zohdi [209]). Such a framework makes the methods previously presented in this monograph applicable.

Remark. There exist a large number of what one can term as "rule-driven" swarms, whereby interaction is not governed by the principles of mechanics but by proximal in-

[25]The interaction "forces" can be, for example, in unmanned airborne vehicles (UAVs), motorized propulsion arising from intervehicle communication.

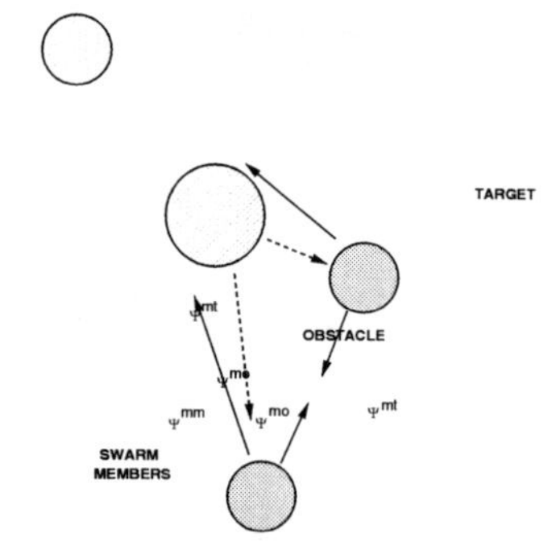

Figure 6.1. *Interaction between the various components (Zohdi* [209]*).*

structions such as, "if a fellow swarm member gets close to me, attempt to retreat as far as possible," "follow the leader," "stay in clusters," etc. While these rule-driven paradigms are usually easy to construct, they are difficult to analyze mathematically. It is primarily for this reason that a mechanical approach is adopted here. Recent broad overviews of the field can be found in Kennedy and Eberhart [120] and Bonabeau et al. [34]. The approach taken is based on work found in Zohdi [209].

6.1 Basic constructions

In the analysis to follow, we treat the swarm members as point masses, i.e., we ignore their dimensions.[26] For each swarm member (N_p in total) the equations of motion are

$$m_i \ddot{\boldsymbol{r}}_i = \boldsymbol{\Psi}^{tot}(\boldsymbol{r}_1, \boldsymbol{r}_2, \ldots, \boldsymbol{r}_{N_p}), \tag{6.1}$$

where $\boldsymbol{\Psi}^{tot}$ represents the forces of interaction between swarm member i and the target, obstacles, and other swarm members. We consider the decomposition (see Figure 6.1)

$$\boldsymbol{\Psi}^{tot} = \boldsymbol{\Psi}^{mm} + \boldsymbol{\Psi}^{mt} + \boldsymbol{\Psi}^{mo}, \tag{6.2}$$

where between swarm members (member-member) we have

$$\boldsymbol{\Psi}^{mm} = \sum_{j \neq i}^{N_p} \left(\left(\underbrace{\alpha_1^{mm} ||\boldsymbol{r}_i - \boldsymbol{r}_j||^{\beta_1^{mm}}}_{\text{attraction}} - \underbrace{\alpha_2^{mm} ||\boldsymbol{r}_i - \boldsymbol{r}_j||^{-\beta_2^{mm}}}_{\text{repulsion}} \right) \underbrace{\frac{\boldsymbol{r}_i - \boldsymbol{r}_j}{||\boldsymbol{r}_i - \boldsymbol{r}_j||}}_{\text{unit vector}} \right), \tag{6.3}$$

where $||\cdot||$ represents the Euclidean norm in R^3, while between the swarm members and the target (member-target) we have

$$\boldsymbol{\Psi}^{mt} = \left(\alpha^{mt} ||\boldsymbol{r}_* - \boldsymbol{r}_i||^{\beta^{mt}} \right) \frac{\boldsymbol{r}_* - \boldsymbol{r}_i}{||\boldsymbol{r}_* - \boldsymbol{r}_i||}, \tag{6.4}$$

[26] The swarm member centers, which are initially nonintersecting, cannot intersect later due to the singular repulsion terms.

and for the repulsion between swarm members and the obstacles (member-obstacle), we have

$$\boldsymbol{\Psi}^{mo} = -\sum_{j=1}^{q}\left(\left(\alpha^{mo}||\boldsymbol{r}_{oj}-\boldsymbol{r}_i||^{-\beta^{mo}}\right)\frac{\boldsymbol{r}_{oj}-\boldsymbol{r}_i}{||\boldsymbol{r}_{oj}-\boldsymbol{r}_i||}\right), \tag{6.5}$$

where q is the number of obstacles and all of the (design) parameters, the α's and β's, are nonnegative.

Remark. One can describe the relative contributions of repulsion and attraction between members of the swarm by considering an individual pair in static equilibrium:

$$\boldsymbol{\Psi}^{mm} = \left(\alpha_1^{mm}||\boldsymbol{r}_i-\boldsymbol{r}_j||^{\beta_1^{mm}} - \alpha_2^{mm}||\boldsymbol{r}_i-\boldsymbol{r}_j||^{-\beta_2^{mm}}\right)\frac{\boldsymbol{r}_i-\boldsymbol{r}_j}{||\boldsymbol{r}_i-\boldsymbol{r}_j||} = \boldsymbol{0}. \tag{6.6}$$

This characterizes a separation length scale describing the tendency to cluster or spread apart:

$$||\boldsymbol{r}_i-\boldsymbol{r}_j|| = \left(\frac{\alpha_2^{mm}}{\alpha_1^{mm}}\right)^{\frac{1}{\beta_1^{mm}+\beta_2^{mm}}} \stackrel{\text{def}}{=} \rho^{mm}. \tag{6.7}$$

We remark that one could have moving targets and obstacles as well as attractive forces between the swarm and the obstacles and repulsive forces from the targets. Adding attractive forces from the obstacles and repulsive forces from the targets makes sense for some applications, for example, in traffic flow, where one does not want the vehicle to hit the target, although we did not consider such cases in the present work.

6.2 A model objective function

As a representative of a class of model problems, we now consider inverse problems whereby the coefficients in the interaction forces are sought, the α's and β's, that deliver desired swarm-like behavior by minimizing a normalized cost function (normalized by the total simulation time and the initial separation distance) representing (1) the time it takes for the swarm members to get to the target and (2) the distance of the swarm members from the target:

$$\Pi = \frac{\left(\int_0^{\mathcal{T}}\sum_{i=1}^{N_p}||\boldsymbol{r}_i-\boldsymbol{r}_*||\,dt\right)}{\mathcal{T}\sum_{i=1}^{N_p}||\boldsymbol{r}_i(t=0)-\boldsymbol{r}_*||}, \tag{6.8}$$

where the total simulation time is $\mathcal{T}=1$; where, for example, for each α, $\alpha^- \le \alpha \le \alpha^+$, and for each β, $\beta^- \le \beta \le \beta^+$; where $\boldsymbol{r}_*$ is the position of the target; and where $\boldsymbol{\alpha}^-$, $\boldsymbol{\alpha}^+$, $\boldsymbol{\beta}^-$, and $\boldsymbol{\beta}^+$ are the lower and upper limit coefficients in the interaction forces. We wish to enforce that, if a swarm member gets too close to an obstacle, it becomes immobilized. Thus, as a side condition, for all t, for all r_{oj}, and for $\tau < \mathcal{T}$, if

$$||\boldsymbol{r}_i(t=\tau)-\boldsymbol{r}_{oj}|| \le R, \tag{6.9}$$

then $\boldsymbol{r}_i = \boldsymbol{r}_i(t=\tau)$ for all $t \ge \tau$, where the unilateral condition represents the effect of being near a "destructive" obstacle. The swarm member is stopped in the position where it enters the "radius of destruction" (R). Therefore, the swarm performance (Π) is severely penalized if it loses members to the obstacles.

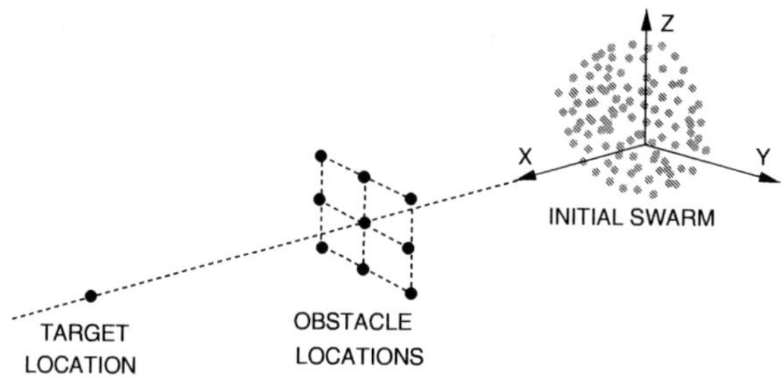

Figure 6.2. *The initial setup for a swarm example (Zohdi* [209]*).*

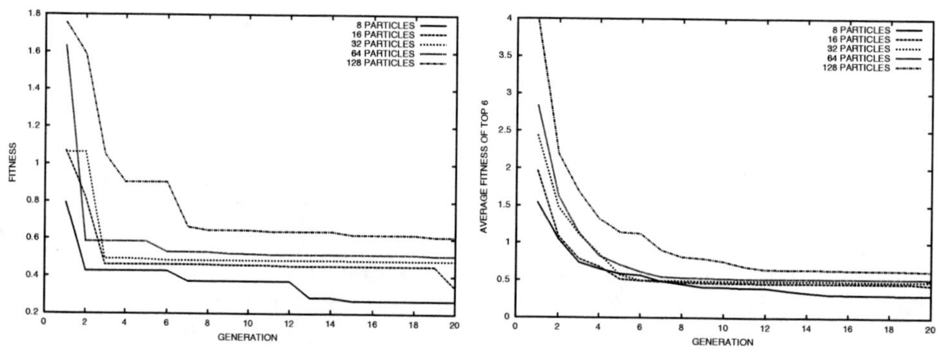

Figure 6.3. *Generational values of (left) the best design's objective function and (right) the average of the best six designs' objective functions for various swarm member sizes (Zohdi* [209]*).*

6.3 Numerical simulation

We consider the situation illustrated in Figure 6.2. The components of the initial position vectors of the nonintersecting swarm members, each assigned a mass[27] of 10 kg, were given random values of $-1 \leq r_{ix}, r_{iy}, r_{iz} \leq 1$. The location of the target was (10, 0, 0). The location of the center of the (rectangular) obstacle array was (5, 0, 0). A nine-obstacle "fence" was set up as follows: (5, 0, 0), (5, 2, 2), (5, 2, −2), (5, 2, 0), (5, 0, 2), (5, −2, −2), (5, −2, 2), (5, −2, 0), (5, 0, −2). The radius of "destruction" for the swarm member-obstacle pair was set to $R = 0.5$. In order to study the effects of the swarm size on the optimal performance, we considered swarms of successively larger sizes, containing $N = 8$, 16, 32, 64, and 128 members. We employed the genetic algorithm introduced in Chapter 5. The search space was, for each α, $10^{-6} \leq \alpha \leq 10^{6}$, and, for each β, $10^{-6} \leq \beta \leq 1$. The number of genetic strings was set to $S = 20$, for $G = 20$ generations, keeping the top six offspring of the top six parents. Therefore, after each generation, eight new genetic strings were introduced. Figure 6.3 depicts the results. The total number of function evaluations of Π is $S + (G-1) \times (S-Q) = 286$, where $G = 20$ is the number of generations, $S = 20$

[27] This is a typical mass of a UAV.

Table 6.1. *The top fitness and average of the top six fitnesses for various swarm sizes (Zohdi* [209]*).*

Swarm Members	Total Strings Tested	Strings Design	Π	$\frac{1}{6}\sum_{i=1}^{6}\Pi_i$
8	1573	5.5000	0.2684	0.3008
16	1646	5.7552	0.3407	0.4375
32	1022	3.5734	0.4816	0.4829
64	1241	4.3391	0.5092	0.5153
128	1970	6.8881	0.6115	0.6210

Table 6.2. *The optimal coefficients of attraction and repulsion for various s warm sizes (Zohdi* [209]*).*

Swarm Members	α_1^{mm}	α_2^{mm}	α^{mt}	α^{mo}
8	451470.44	270188.87	735534.64	141859.99
16	128497.49	279918.51	778117.81	80526.85
32	111642.28	564292.53	872627.48	7899.69
64	394344.61	625999.39	910734.12	23961.73
128	767084.35	264380.23	574909.53	159249.40

Table 6.3. *The optimal exponents of attraction and repulsion for various swarm sizes (Zohdi* [209]*).*

Swarm Members	β_1^{mm}	β_2^{mm}	β^{mt}	β^{mo}
8	0.8555	0.2686	0.4366	0.6433
16	0.1793	0.1564	0.8101	0.8386
32	0.4101	0.0404	0.7995	0.5632
64	0.4030	0.1148	0.7422	0.4976
128	0.5913	0.0788	0.5729	0.8313

Table 6.4. *The ratios of optimal repulsion and attraction for various swarm sizes (Zohdi* [209]*).*

Swarm Members	ρ^{mm}
8	0.6333
16	10.1622
32	36.4685
64	2.4407
128	0.2040

is the total number of genetic strings in the population, and $Q = 6$ is the number of parents kept after each generation. The total time was set (normalized) to be one second ($\mathcal{T} = 1$).

From Tables 6.1–6.4, there appears to be no convergence in the optima with respect to the swarm member number. A clear result is that one cannot expect optima for one swarm size to be optimal for another. In other words, there is no apparent scaling law. In Figures 6.4–6.6, frames are shown for the 128-particle swarm. The 128-particle swarm bunches up and moves through the obstacle fence unharmed (centered at (5, 0, 0)) by going underneath the central

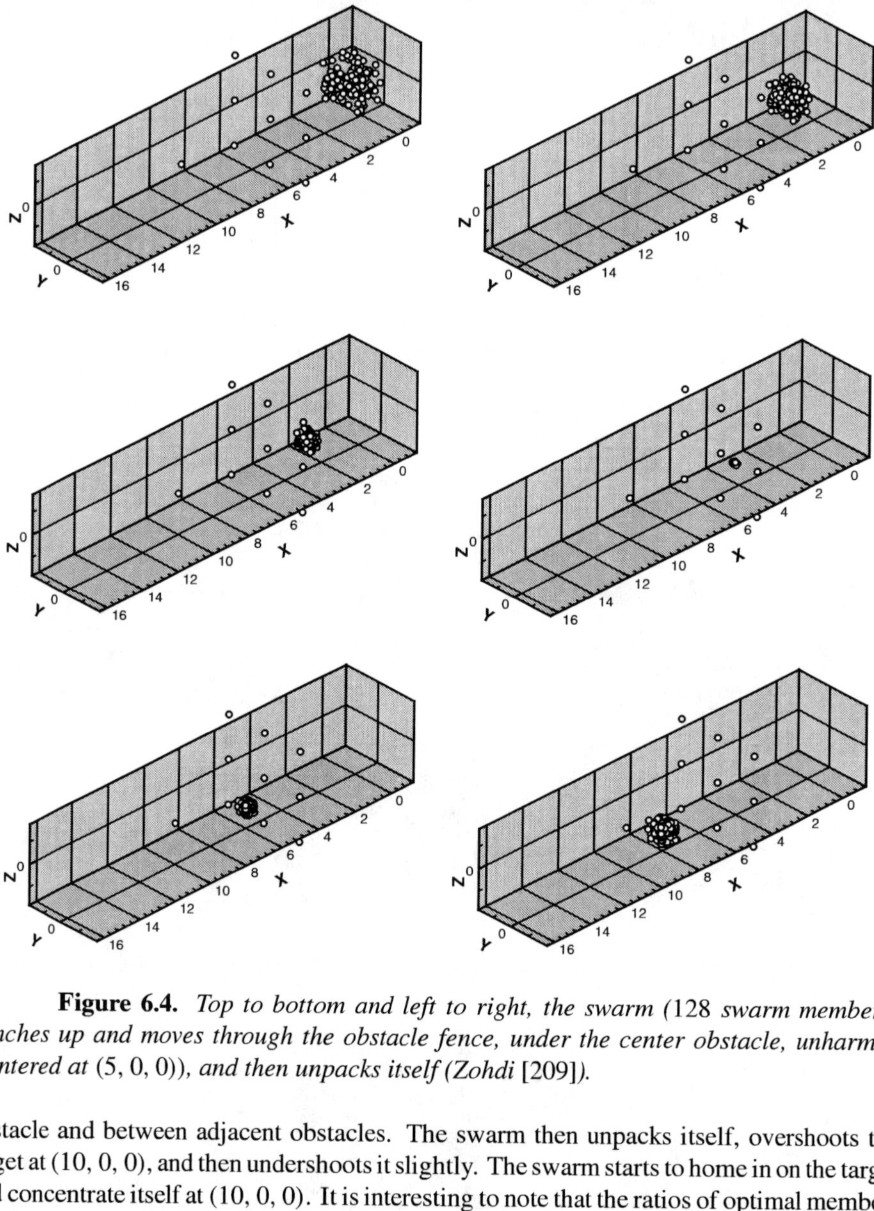

Figure 6.4. *Top to bottom and left to right, the swarm* (128 *swarm members*) *bunches up and moves through the obstacle fence, under the center obstacle, unharmed* (*centered at* (5, 0, 0)), *and then unpacks itself* (*Zohdi* [209]).

obstacle and between adjacent obstacles. The swarm then unpacks itself, overshoots the target at (10, 0, 0), and then undershoots it slightly. The swarm starts to home in on the target and concentrate itself at (10, 0, 0). It is interesting to note that the ratios of optimal member-member repulsion to attraction (ρ^{mm}) are quite small for the 128-particle swarm; however, for other swarm sizes, such as 16 and 32, the optima are relatively large. This implies that bunching up is not necessarily the best strategy to surround the target for every swarm size.

6.4 Discussion

In many applications, the computed positions, velocities, and accelerations of the members of a swarm, for example, people or vehicles, must be translated into realizable movement.

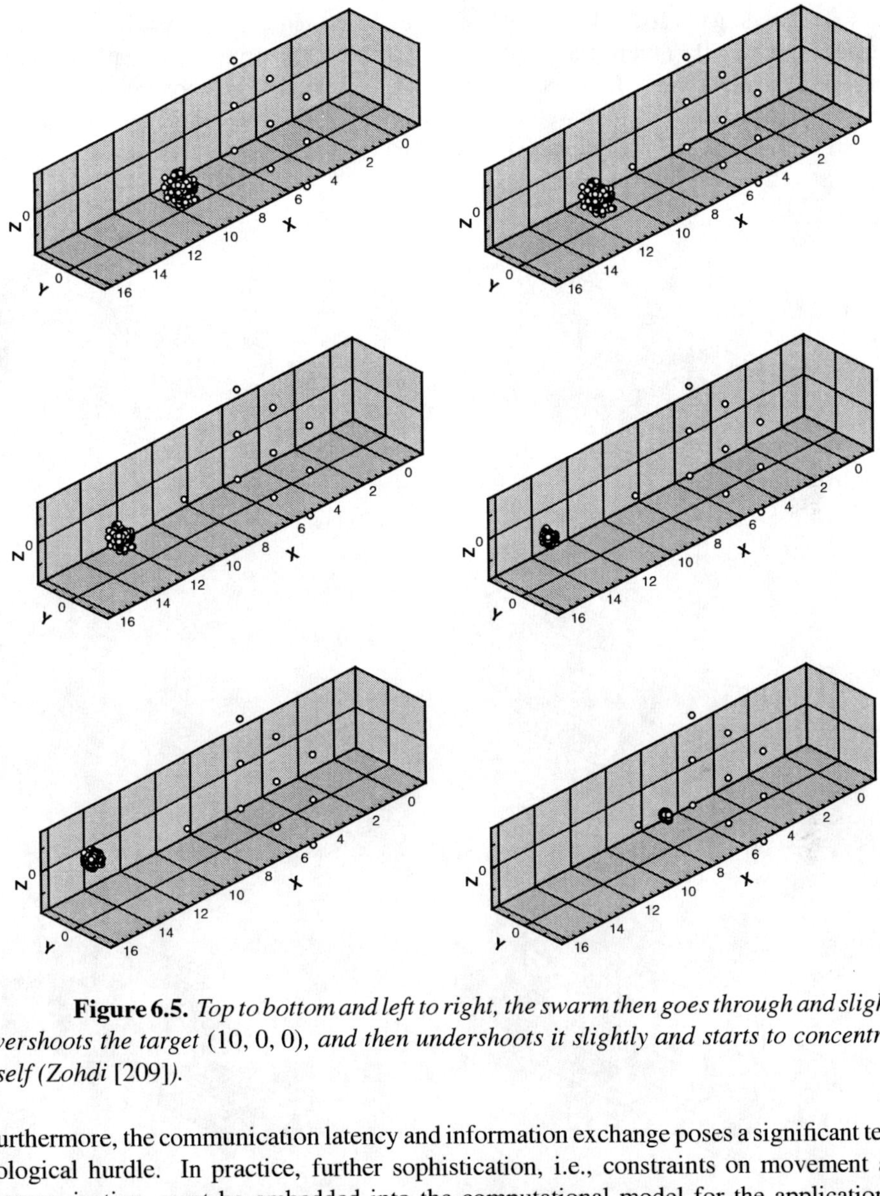

Figure 6.5. *Top to bottom and left to right, the swarm then goes through and slightly overshoots the target* (10, 0, 0)*, and then undershoots it slightly and starts to concentrate itself (Zohdi* [209]*).*

Furthermore, the communication latency and information exchange poses a significant technological hurdle. In practice, further sophistication, i.e., constraints on movement and communication, must be embedded into the computational model for the application at hand. However, the fundamental computational philosophy and modeling strategy should remain relatively unchanged. It is important to remark on a fundamental set of results found in Hedrick and Swaroop [92], Hedrick et al. [93], Swaroop and Hedrick [183], [184], and Shamma [175], namely, that if the interaction is only with the nearest neighbors, and if there is no inertial reference point for the swarm members to refer to, instabilities (collisions) may occur. In the present analysis, such inertial reference points were furnished by the fact that the members of the swarm knew the absolute locations of the stationary obstacles and target. Also, because the communication for a given swarm member was with all other members, the stability was a nonissue. Furthermore, due to the presence of a $\frac{1}{r}$-type interaction force between the initially nonoverlapping swarm members, the centers could not intersect (a singular repulsion term). However, if the target and obstacles begin to move in response to the swarm, which may be the case in certain applications, and the communication between swarm members is only with the nearest neighbors (a possible technological restriction), then instabilities can become a primary concern.

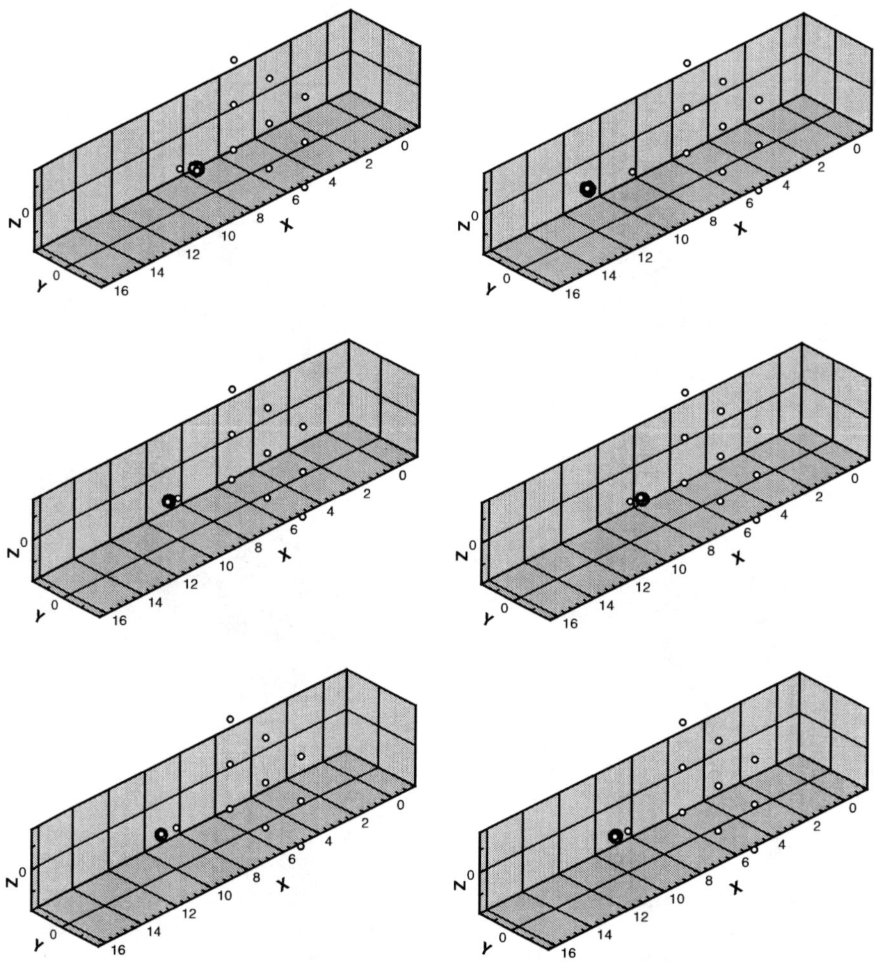

Figure 6.6. *Top to bottom and left to right, the swarm starts to oscillate slightly around the target and then begins to home in on the target and concentrate itself at* (10, 0, 0) *(Zohdi* [209]*).*

Chapter 7

Advanced particulate flow models

We now return to the issue of particulate flows. In many applications, emphasis is placed on describing possible particle clustering, which can lead to the formation of larger structures within the particulate flow. This requires slight modification of the potentials introduced earlier. The approach in this chapter draws from general methods developed in Zohdi [217].

7.1 Introduction

There has been a steady increase in analysis of complex particulate flows, where multifield phenomena, such as electrostatic charging and thermochemical coupling, are of interest. Such systems arise in the study of clustering and aggregation of particles in natural science applications where particles collide, cluster, and grow into larger objects. Understanding coupled phenomena in particulate flows is also of interest in modern industrial processes that involve spray processes such as epitaxy and sputtering as well as dust control, etc. For example, in many processes, intentional charging and heating of particulates, such as those in inkjet printers, is critical. Thus, in addition to the calculation of the dynamics of the particles in the particulate flow, thermal fields must be determined simultaneously to be able to make accurate predictions of the behavior of the flow. Accordingly, the present work develops models and robust solution strategies to perform direct simulation of the dynamics of particulate media in the presence of thermal effects.

7.2 Clustering and agglomeration via binding forces

In many applications, the near-fields can dramatically change when the particles are very close to one another, leading to increased repulsion or attraction. Of specific interest in this work is interparticle binding leading to clustering and agglomeration (Figure 7.1). A particularly easy way to model this is via a near-field attractive augmentation of the form

$$\underbrace{\boldsymbol{\Psi}_i}_{\text{UNAUGMENTED}} + \underbrace{\alpha_a ||\boldsymbol{r}_i - \boldsymbol{r}_j||^{-\beta_a} \boldsymbol{n}_{ij}}_{\boldsymbol{\Psi}^a \stackrel{\text{def}}{=} \text{BINDING FORCE (AUGMENTATION)}}, \tag{7.1}$$

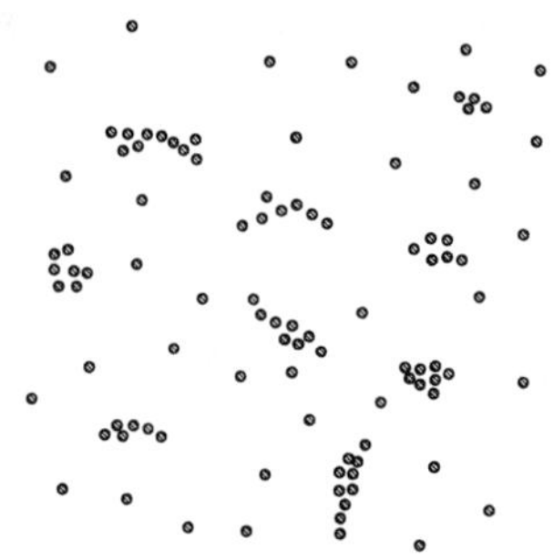

Figure 7.1. *Clustering within a particulate flow (Zohdi* [217]*).*

which is activated if

$$||\boldsymbol{r}_i - \boldsymbol{r}_j|| \leq (b_i + b_j)\delta_a, \tag{7.2}$$

where b_i and b_j are the radii of the particles[28] and $1 \leq \delta_a$ is the critical distance needed for the augmentation to become active. The corresponding binding potential is

$$V_a(||\boldsymbol{r}_i - \boldsymbol{r}_j||) = \frac{\alpha_a ||\boldsymbol{r}_i - \boldsymbol{r}_j||^{-\beta_a+1}}{-\beta_a + 1}, \tag{7.3}$$

which is active if $||\boldsymbol{r}_i - \boldsymbol{r}_j|| \leq (b_i + b_j)\delta_a$. Denoting the nominal (unagglomerated) equilibrium distance by d_e and the equilibrium distance when agglomeration is active by d_a, we have, with $\beta_a = \beta_1$,

$$||\boldsymbol{r}_i - \boldsymbol{r}_j|| = \left(\frac{\alpha_2}{\alpha_1 + \alpha_a}\right)^{\frac{1}{-\beta_1+\beta_2}} = d_a \leq d_e = \left(\frac{\alpha_2}{\alpha_1}\right)^{\frac{1}{-\beta_1+\beta_2}}. \tag{7.4}$$

Clearly, with such a model, the magnitude of α_a must be limited so that no interpenetration of the particles is possible, i.e., $||\boldsymbol{r}_i - \boldsymbol{r}_j|| \geq b_i + b_j$ must hold at all times.

Remark. For many engineering materials, some surface adhesion persists, which can lead to a sticking phenomenon between surfaces, even when no explicit charging has occurred. For more details, see Tabor [186] and, specifically for "clumping," see the book by Rietema [170].

7.3 Long-range instabilities and interaction truncation

Let us reconsider the dynamics of the particle in the (one-dimensional) normal direction, with a perturbation

$$\tilde{r} = r + \delta r, \tag{7.5}$$

leading to

$$m\ddot{\tilde{r}} = \Psi^{nf}(\tilde{r}), \tag{7.6}$$

where r is the perturbation-free position vector of the particle, governed by

$$m\ddot{r} = \Psi^{nf}(r). \tag{7.7}$$

[28]They will be taken to be the same later in the simulations.

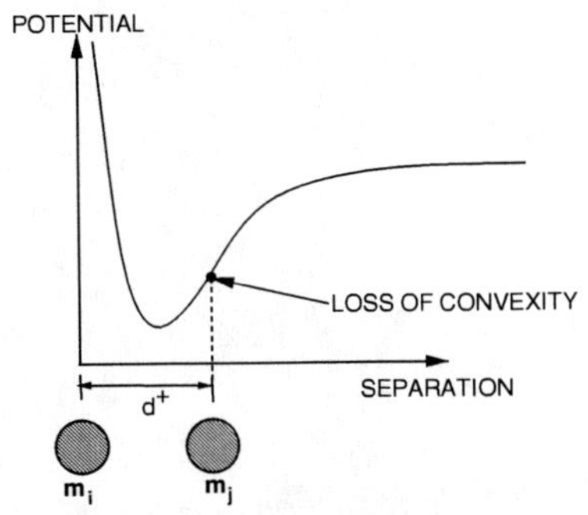

Figure 7.2. *Identification of an inflection point (loss of convexity (Zohdi [217])).*

Subtracting Equation (7.6) from Equation (7.7), we have

$$m\delta\ddot{\tilde{r}} = \Psi^{nf}(\tilde{r}) - \Psi^{nf}(r) \approx \frac{\partial \Psi^{nf}}{\partial r}|_{\tilde{r}=r}\delta r + \cdots, \tag{7.8}$$

resulting in

$$m\delta\ddot{\tilde{r}} \approx \left.\frac{\partial \Psi^{nf}}{\partial r}\right|_{\tilde{r}=r}\delta r \Rightarrow m\delta\ddot{\tilde{r}} - \left.\frac{\partial \Psi^{nf}}{\partial r}\right|_{\tilde{r}=r}\delta r \approx 0. \tag{7.9}$$

If $\frac{\partial \Psi^{nf}(r)}{\partial r}$ is positive, there will be exponential growth of the perturbation, while if $\frac{\partial \Psi^{nf}(r)}{\partial r}$ is negative, there will be oscillatory behavior of the perturbation. Thus, since

$$-\frac{\partial^2 V}{\partial r^2} = \frac{\partial \Psi^{nf}}{\partial r}, \tag{7.10}$$

we have

$$m\delta\ddot{\tilde{r}} + \frac{\partial^2 V}{\partial r^2}|_{\tilde{r}=r}\delta r \approx 0. \tag{7.11}$$

Thus, for stability, the potential should be convex about r. Clearly, the point at which the potential changes from a convex to a concave character is the point of long-range instability (Figure 7.2).[29] For motion in the normal direction, we have

$$\frac{\partial^2 V}{\partial r^2} = -\beta_1\alpha_1|r - r_o|^{-\beta_1-1} + \beta_2\alpha_2|r - r_o|^{-\beta_2-1} = 0, \tag{7.12}$$

thus leading to

$$|r - r_o| = \left(\frac{\beta_2\alpha_2}{\beta_1\alpha_1}\right)^{\frac{1}{-\beta_1+\beta_2}} = d^{(+)}. \tag{7.13}$$

[29] As mentioned before, for the central force potential form chosen in this work, it suffices to study the motion in the normal direction, i.e., the line connecting the centers of the particles. For disturbances in directions orthogonal to the normal direction, the potential is neutrally stable, i.e., the Hessian's determinant is zero, thus indicating that the potential does not change for such perturbations.

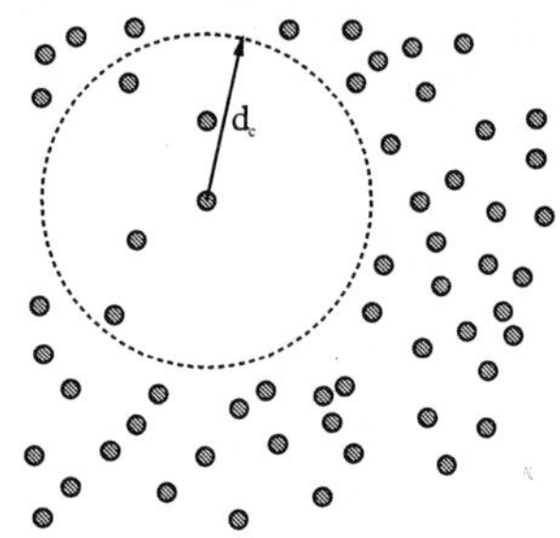

Figure 7.3. *Introduction of a cutoff function.*

Thus, the preceding analysis indicates that, for the three-dimensional case, an interaction "cutoff" distance (d_c) should be introduced (Figure 7.3),

$$||\boldsymbol{r}_i - \boldsymbol{r}_j|| = d_c \leq d^{(+)}, \tag{7.14}$$

to avoid long-range (central-force) instabilities.

Remark. By introducing a cutoff distance, one can circumvent a loss-of-convexity instability. However, introducing such a cutoff can induce another type of instability. Specifically, if the particles are in static equilibrium, or are not approaching one another, and if the cutoff distance, d_c, is much smaller than the static equilibrium separation distance, d_e, then the particles will not interact at all. Thus, we have the following two-sided bounds on the cutoff for near-field forces to play a physically realistic role:

$$\left(\frac{\alpha_2}{\alpha_1}\right)^{\frac{1}{-\beta_1+\beta_2}} = d^{(-)} \leq d_c \leq d^{(+)} = \left(\frac{\beta_2\alpha_2}{\beta_1\alpha_1}\right)^{\frac{1}{-\beta_1+\beta_2}}. \tag{7.15}$$

Clearly, since $\beta_2 > \beta_1, d^{(-)}$ is a lower bound (dictated by the minimum interaction distance), while $d^{(+)}$ is an upper bound (dictated by the (convexity-type) stability).

7.4 A simple model for thermochemical coupling

As indicated earlier, in certain applications, in addition to the near-field and contact effects introduced thus far, thermal behavior is of interest. For example, applications arise in the study of interstellar particulate dust flows in the presence of dilute hydrogen-rich gas. In many cases, the source of heat generated during impact in such flows can be traced to the reactivity of the particles. This affects the mechanics of impact, for example, due to thermal softening. For instance, the presence of a reactive substance (gas) adsorbed onto the surface of interplanetary dust can be a source of intense heat generation, through thermochemical reactions activated by impact forces, which thermally softens the material, thus reducing the coefficient of restitution, which in turn strongly affects the mechanical impact event itself (Figure 7.4).

To illustrate how one can incorporate thermal effects, a somewhat ad hoc approach, building on the relation in Equation (2.50), is to construct a thermally dependent coefficient

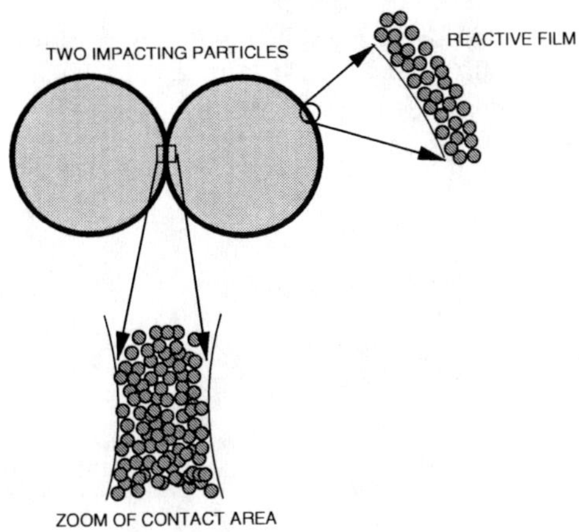

Figure 7.4. *Presence of dilute (smaller-scale) reactive gas particles adsorbed onto the surface of two impacting particles (Zohdi* [217]*).*

of restitution as follows (multiplicative decomposition):

$$e \stackrel{\text{def}}{=} \left(\max\left(e_o\left(1-\frac{\Delta v_n}{v^*}\right), e^-\right)\right)\left(\max\left(\left(1-\frac{\theta}{\theta^*}\right), 0\right)\right), \tag{7.16}$$

where θ^* can be considered as a thermal softening temperature. In order to determine the thermal state of the particles, we shall decompose the heat generation and heat transfer processes into two stages. Stage I describes the extremely short time interval when impact occurs, $\delta t \ll \Delta t$, and accounts for the effects of chemical reactions, which are relevant in certain applications, and energy release due to mechanical straining. Stage II accounts for the postimpact behavior involving convective and radiative effects.

7.4.1 Stage I: An energy balance during impact

Throughout the analysis, we shall use extremely simple, basic, models. Consistent with the particle-based philosophy, it is assumed that the temperature fields are uniform in the particles.[30] We consider an energy balance, governing the interconversions of mechanical, thermal, and chemical energy in a system, dictated by the first law of thermodynamics. Accordingly, we require the time rate of change of the sum of the kinetic energy ($\mathcal{K}$) and the stored energy ($\mathcal{S}$) to be equal to the sum of the work rate (power, $\mathcal{P}$) and the net heat supplied ($\mathcal{H}$):

$$\frac{d}{dt}(\mathcal{K}+\mathcal{S}) = \mathcal{P}+\mathcal{H}, \tag{7.17}$$

where the stored energy comprises a thermal part,

$$\mathcal{S} = mC\theta, \tag{7.18}$$

---

[30]Thus, the gradient of the temperature within the particle is zero, i.e., $\nabla\theta = \mathbf{0}$. Thus, a Fourier-type law for the heat flux will register a zero value, $\boldsymbol{q} = -\mathbb{K}\cdot\nabla\theta = \mathbf{0}$.

where C is the heat capacity per unit mass and, consistent with our assumptions that the particles deform negligibly during impact, we assume that there is an *insignificant amount of mechanically stored energy*. The kinetic energy is

$$\mathcal{K} = \frac{1}{2} m \boldsymbol{v} \cdot \boldsymbol{v}. \tag{7.19}$$

The mechanical power term is due to the forces acting on a particle, namely

$$\mathcal{P} = \frac{d\mathcal{W}}{dt} = \boldsymbol{\Psi}^{tot} \cdot \boldsymbol{v}, \tag{7.20}$$

and, because

$$\frac{d\mathcal{K}}{dt} = m\dot{\boldsymbol{v}} \cdot \boldsymbol{v}, \tag{7.21}$$

and we have a balance of momentum

$$m\dot{\boldsymbol{v}} \cdot \boldsymbol{v} = \boldsymbol{\Psi}^{tot} \cdot \boldsymbol{v}, \tag{7.22}$$

we have

$$\frac{d\mathcal{K}}{dt} = \frac{d\mathcal{W}}{dt} = \mathcal{P}, \tag{7.23}$$

leading to

$$\frac{d\mathcal{S}}{dt} = \mathcal{H}. \tag{7.24}$$

For example, in certain applications of interest, such as the ones mentioned, we consider that the primary source of heat is due to chemical reactions, where the reactive layer generates heat upon impact. The chemical reaction energy is defined as

$$\delta\mathcal{H} \stackrel{\text{def}}{=} \int_t^{t+\delta t} \mathcal{H}\, dt. \tag{7.25}$$

Equation (7.24) can be rewritten for the temperature at time $= t + \delta t$ as

$$\theta(t + \delta t) = \theta(t) + \frac{\delta\mathcal{H}}{mC}. \tag{7.26}$$

The energy released from the reactions is assumed to be proportional to the amount of the gaseous substance available to be compressed in the contact area between the particles. A typical ad hoc approximation in combustion processes is to write, for example, a linear relation

$$\delta\mathcal{H} \approx \kappa \min\left(\frac{|\overline{I}_n|}{I_n^*}, 1\right) \pi b^2, \tag{7.27}$$

where $\overline{I}_n$ is the normal impact force; κ is a reaction (saturation) constant, energy per unit area; I_n^* is a normalization parameter; and b is the particle radius. For details, see Schmidt [172], for example. For the grain sizes and material properties of interest, the term in Equation (7.26), $\frac{\delta\mathcal{H}}{mC}$, indicates that values of approximately $\kappa \approx 10^6$ J/m^2 will generate significant amounts of heat.[31] *Clearly, these equations are coupled to those of impact through the coefficient of restitution and the velocity-dependent impulse.* Additionally, the postcollision velocities are computed from the momentum relations coupled to the temperature. Later in the analysis, this equation is incorporated into an overall staggered fixed-point iteration scheme, whereby the temperature is predicted for a given velocity field, and then the velocities are recomputed with the new temperature field, etc. The process is repeated until the fields change negligibly between successive iterations. The entire set of equations are embedded within a larger overall set of equations later in the analysis and are solved in a recursively staggered manner.

7.4.2 Stage II: Postcollision thermal behavior

After impact, it is assumed that a process of convection, for example, governed by Newton's law of cooling, and radiation, according to a simple Stefan–Boltzmann law, occurs. As before, it is assumed that the temperature fields are uniform within the particles, so conduction within the particles is negligible. We remark that the validity of using a lumped thermal model, i.e., ignoring temperature gradients and assuming a uniform temperature within a particle, is dictated by the magnitude of the Biot number. A small Biot number indicates that such an approximation is reasonable. The Biot number for spheres scales with the ratio of the particle volume (V) to the particle surface area (a_s), $\frac{V}{a_s} = \frac{b}{3}$, which indicates that a uniform temperature distribution is appropriate, since the particles, by definition, are small. We also assume that the dynamics of the (dilute) gas does not affect the motion of the (much heavier) particles. The gas only supplies a reactive thin film on the particles' surfaces. The first law reads

$$\frac{d(K+U)}{dt} = m\dot{\boldsymbol{v}} \cdot \boldsymbol{v} + mC\dot{\theta} = \underbrace{\boldsymbol{\Psi}^{tot} \cdot \boldsymbol{v}}_{mechanical\ power} \underbrace{- h_c a_s(\theta - \theta_o)}_{convective\ heating} \underbrace{- \mathcal{B} a_s \epsilon (\theta^4 - \theta_s^4)}_{far\text{-}field\ radiation}, \quad (7.28)$$

where θ_o is the temperature of the ambient gas; θ_s is the temperature of the far-field surface (for example, a container surrounding the flow) with which radiative exchange is made; $\mathcal{B} = 5.67 \times 10^{-8} \frac{W}{m^2-K}$ is the Stefan–Boltzmann constant; $0 \le \epsilon \le 1$ is the emissivity, which indicates how efficiently the surface radiates energy compared to a black-body (an ideal emitter); $0 \le h_c$ is the heating due to convection (Newton's law of cooling) into the dilute gas; and a_s is the surface area of a particle. It is assumed that the radiation exchange between the particles (emission exchange between particles) is negligible.[32] For the applications considered here, typically, h_c is quite small and plays a small role in the heat transfer processes.[33] From a balance of momentum, we have $m\dot{\boldsymbol{v}} \cdot \boldsymbol{v} = \boldsymbol{\Psi}^{tot} \cdot \boldsymbol{v}$, and Equation (7.28) becomes

$$mC\dot{\theta} = -h_c a_s(\theta - \theta_o) - \mathcal{B} a_s \epsilon (\theta^4 - \theta_s^4). \quad (7.29)$$

[31] By construction, this model has increased heat production, via $\delta\mathcal{H}$, for increasing κ.

[32] Various arguments for such an assumption can be found in the classical text of Bohren and Huffman [33].

[33] The Reynolds number, which measures the ratio of the inertial forces to viscous forces in the surrounding gas and dictates the magnitude of these parameters, is extremely small in the regimes considered.

Therefore, after temporal integration with the previously used finite difference time step of $\Delta t \gg \delta t$, we have[34]

$$\theta(t+\Delta t) = \frac{mC}{mC + h_c a_s \Delta t}\bar{\theta}(t) - \frac{\Delta t \mathcal{B} a_s \epsilon}{mC + h_c a_s \Delta t}\left(\theta^4(t+\Delta t) - \theta_s^4\right) + \frac{h_c a_s \Delta t \theta_o}{mC + h_c a_s \Delta t}, \tag{7.30}$$

where $\bar{\theta}(t) \stackrel{\text{def}}{=} \theta(t+\delta t)$ is computed from Equation (7.26). This implicit nonlinear equation for $\theta(t + \Delta t)$, *for each particle*, is solved simultaneously with the equations for the dynamics of the particles by employing a *multifield* staggering scheme, which we shall discuss momentarily.

Remark. Convection heat transfer comprises two primary mechanisms, one due to primarily random molecular motion (diffusion) and the other due to bulk motion of a fluid, in our case a gas, surrounding the particles. As we have indicated, in the applications of interest, the gas is dilute and the Reynolds number is small, so convection plays a very small role in the heat transfer process for dry particulate flows in the presence of a dilute gas. The nondilute surrounding fluid case will be considered in Chapter 8. Also, we recall that a black-body is an ideal radiating surface with the following properties:

- A black-body absorbs all incident radiation, regardless of wavelength and direction.
- For a prescribed temperature and wavelength, no surface can emit more energy than a black-body.
- Although the radiation emitted by a black-body is a function of wavelength and temperature, it is independent of direction.

Since a black-body is a perfect emitter, it serves as a standard against which the radiative properties of actual surfaces may be compared. The Stefan–Boltzmann law, which is computed by integrating the Planck representation of the emissive power distribution of a black-body over all wavelengths,[35] allows the calculation of the amount of radiation emitted in all directions and over all wavelengths simply from the knowledge of the temperature of the black-body. We note that Equation (7.30) is of the form

$$\theta(t+\Delta t) = \mathcal{G}(\theta(t+\Delta t)) + \mathcal{R}, \tag{7.31}$$

where $\mathcal{R} \neq \mathcal{R}(\theta(t+\Delta t))$ and $\mathcal{G}$'s behavior is controlled by

$$\frac{\Delta t \mathcal{B} a_s \epsilon}{mC + h_c a_s \Delta t}, \tag{7.32}$$

which is quite small. Thus, a fixed-point iterative scheme such as

$$\theta^K(t+\Delta t) = \mathcal{G}(\theta^{K-1}(t+\Delta t)) + \mathcal{R} \tag{7.33}$$

would converge rapidly.

[34]For this stage, since $\delta t \ll \Delta t$, we assign $\overline{\theta}(t) = \theta(t+\delta t) = \theta(t) + \frac{\delta \mathcal{H}}{mC}$ and replace $\theta(t)$ with it in Equation (7.30).

[35]Radiation is idealized as requiring no medium to transmit energy.

7.5 Staggering schemes

Broadly speaking, staggering schemes proceed by solving each field equation individually, allowing only the primary field variable to be active. After the solution of each field equation, the primary field variable is updated, and the next field equation is addressed in a similar manner. Such approaches have a long history in the computational mechanics community. For example, see Park and Felippa [161], Zienkiewicz [206], Schrefler [173], Lewis et al. [133], Doltsinis [52], [53], Piperno [162], Lewis and Schrefler [132], Armero and Simo [7]–[9], Armero [10], Le Tallec and Mouro [131], Zohdi [208], [209], and the extensive works of Farhat and coworkers (Piperno et al. [163], Farhat et al. [65], Lesoinne and Farhat [130], Farhat and Lesoinne [66], Piperno and Farhat [163], and Farhat et al. [67]). Generally speaking, if a recursive staggering process is *not employed* (an explicit scheme), the staggering error can accumulate rapidly. However, an overkill approach involving very small time steps, smaller than needed to control the discretization error, simply to suppress a nonrecursive staggering process error, is computationally inefficient. Therefore, the objective of the next section is to develop a strategy to adaptively adjust, in fact maximize, the choice of the time step size to control the staggering error, while simultaneously staying below the critical time step size needed to control the discretization error. An important related issue is to simultaneously minimize the computational effort involved. The number of times the multifield system is solved, as opposed to time steps, is taken as the measure of computational effort, since within a time step, many multifield system re-solves can take place. We now develop a staggering scheme by following an approach found in Zohdi [208]–[210].

7.5.1 A general iterative framework

We consider an abstract setting, whereby one solves for the particle positions, assuming the thermal fields fixed,

$$\mathcal{A}_1(\underline{\boldsymbol{r}^{L+1,K}}, \theta^{L+1,K-1}) = \mathcal{F}_1(\boldsymbol{r}^{L+1,K-1}, \theta^{L+1,K-1}), \tag{7.34}$$

and then one solves for the thermal fields, assuming the particle positions fixed,

$$\mathcal{A}_2(\boldsymbol{r}^{L+1,K}, \underline{\theta^{L+1,K}}) = \mathcal{F}_2(\boldsymbol{r}^{L+1,K}, \theta^{L+1,K-1}), \tag{7.35}$$

where only the underlined variable is "active," L indicates the time step, and K indicates the iteration counter. Within the staggering scheme, implicit time-stepping methods, with time step size adaptivity, will be used throughout the upcoming analysis.

Continuing where Equation (3.28) left off, we define the normalized errors within each time step, for the two fields, as

$$\varpi_{rK} \stackrel{\text{def}}{=} \frac{||\boldsymbol{r}^{L+1,K} - \boldsymbol{r}^{L+1,K-1}||}{||\boldsymbol{r}^{L+1,K} - \boldsymbol{r}^{L}||} \quad \text{and} \quad \varpi_{\theta K} \stackrel{\text{def}}{=} \frac{||\theta^{L+1,K} - \theta^{L+1,K-1}||}{||\theta^{L+1,K} - \theta^{L}||}. \tag{7.36}$$

We define the maximum "violation ratio," i.e., the larger of the ratios of each field variable's error to its corresponding tolerance, by $Z_K \stackrel{\text{def}}{=} \max(z_{rK}, z_{\theta K})$, where

$$z_{rK} \stackrel{\text{def}}{=} \frac{\varpi_{rK}}{TOL_r} \quad \text{and} \quad z_{\theta K} \stackrel{\text{def}}{=} \frac{\varpi_{\theta K}}{TOL_\theta}, \tag{7.37}$$

with the minimum scaling factor defined as $\Phi_K \stackrel{\text{def}}{=} \min(\phi_{rK}, \phi_{\theta K})$, where

$$\phi_{rK} \stackrel{\text{def}}{=} \left(\frac{\left(\frac{TOL_r}{\varpi_{r0}}\right)^{\frac{1}{pK_d}}}{\left(\frac{\varpi_{rK}}{\varpi_{r0}}\right)^{\frac{1}{pK}}} \right), \qquad \phi_{\theta K} \stackrel{\text{def}}{=} \left(\frac{\left(\frac{TOL_\theta}{\varpi_{\theta 0}}\right)^{\frac{1}{pK_d}}}{\left(\frac{\varpi_{\theta K}}{\varpi_{\theta 0}}\right)^{\frac{1}{pK}}} \right). \tag{7.38}$$

See Algorithm 7.1. The overall goal is to deliver solutions where the staggering (incomplete coupling) error is controlled and the temporal discretization accuracy dictates the upper limits on the time step size (Δt^{lim}).

Remark. As in the single-field multiple-particle discussion earlier, an alternative approach is to attempt to solve the entire multifield system simultaneously (monolithically). This would involve the use of a Newton-type scheme, which can also be considered as a type of fixed-point iteration. Newton's method is covered as a special case of this general analysis. To see this, let

$$\boldsymbol{w} = (\boldsymbol{r}, \theta), \tag{7.39}$$

and consider the residual defined by

$$\boldsymbol{\Pi} \stackrel{\text{def}}{=} \mathcal{A}(\boldsymbol{w}) - \mathcal{F}. \tag{7.40}$$

Linearization leads to

$$\boldsymbol{\Pi}(\boldsymbol{w}^K) = \boldsymbol{\Pi}(\boldsymbol{w}^{K-1}) + \nabla_{\boldsymbol{w}}\boldsymbol{\Pi}|_{\boldsymbol{w}^{K-1}}(\boldsymbol{w}^K - \boldsymbol{w}^{K-1}) + \mathcal{O}(||\Delta\boldsymbol{w}||^2), \tag{7.41}$$

and thus the Newton updating scheme can be developed by enforcing

$$\boldsymbol{\Pi}(\boldsymbol{w}^K) \approx \boldsymbol{0}, \tag{7.42}$$

leading to

$$\boldsymbol{w}^K = \boldsymbol{w}^{K-1} - (\mathcal{A}^{TAN,K-1})^{-1}\boldsymbol{\Pi}(\boldsymbol{w}^{K-1}), \tag{7.43}$$

where

$$\mathcal{A}^{TAN,K} = (\nabla_{\boldsymbol{w}}\mathcal{A}(\boldsymbol{w}))\,|_{\boldsymbol{w}^K} = (\nabla_{\boldsymbol{w}}\boldsymbol{\Pi}(\boldsymbol{w}))\,|_{\boldsymbol{w}^K} \tag{7.44}$$

is the tangent. Therefore, in the fixed-point form, one has the operator

$$\mathcal{G}(\boldsymbol{w}) = \boldsymbol{w} - (\mathcal{A}^{TAN})^{-1}\boldsymbol{\Pi}(\boldsymbol{w}). \tag{7.45}$$

One immediately sees a fundamental difficulty due to the possibility of a zero or near-zero tangent when employing a Newton's method on a nonconvex system, which can lose positive definiteness and which in turn will lead to an indefinite system of algebraic equations.[36] Therefore, while Newton's method usually converges at a faster rate than a direct fixed-point iteration, quadratically as opposed to superlinearly, its convergence criteria are less robust than the presented fixed-point algorithm, due to its dependence on the gradients of the solution. Furthermore, for the problems considered, the solutions are nonsmooth and nonconvex, primarily due to the impact events, and thus we opted for the more robust "gradientless" staggering scheme.

[36]Furthermore, the tangent may not exist in some (nonsmooth) cases.

(1) GLOBAL FIXED-POINT ITERATION (SET $i = 1$ AND $K = 0$):

(2) IF $i > N_p$, THEN GO TO (4);

(3) IF $i \leq N_p$, THEN (FOR PARTICLE i)

(a) COMPUTE POSITION: $\boldsymbol{r}_i^{L+1,K} \approx \frac{\Delta t^2}{m}\left(\boldsymbol{\Psi}^{tot}(\boldsymbol{r}^{L+1,K-1})\right) + \boldsymbol{r}_i^L + \Delta t \dot{\boldsymbol{r}}_i^L$;

(b) COMPUTE TEMPERATURE (FOR PARTICLE i):

$$\theta_i^{L+1,K} = \theta_i^L + \frac{\delta\mathcal{H}^{L+1,K-1}}{mC};$$

$$\theta_i^{L+1,K} = \frac{mC}{mC + h_c a_s \Delta t}\theta_i^{L+1,K-1} - \frac{\Delta t \mathcal{B} a_s \epsilon}{mC + h_c a_s \Delta t}\left((\theta_i^{L+1,K-1})^4 - \theta_s^4\right)$$

$$+ \frac{h_c a_s \Delta t \theta_o}{mC + h_c a_s \Delta t};$$

(c) GO TO (2) AND NEXT PARTICLE ($i = i + 1$);

(4) ERROR MEASURES (normalized):

(a) $\varpi_{rK} \stackrel{\text{def}}{=} \frac{\sum_{i=1}^{N_p} ||\boldsymbol{r}_i^{L+1,K} - \boldsymbol{r}_i^{L+1,K-1}||}{\sum_{i=1}^{N_p} ||\boldsymbol{r}_i^{L+1,K} - \boldsymbol{r}_i^L||}, \quad \varpi_{\theta K} \stackrel{\text{def}}{=} \frac{\sum_{i=1}^{N_p} ||\theta_i^{L+1,K} - \theta_i^{L+1,K-1}||}{\sum_{i=1}^{N_p} ||\theta_i^{L+1,K} - \theta_i^L||};$

(b) $Z_K \stackrel{\text{def}}{=} \max(z_{rK}, z_{\theta K})$ where $z_{rK} \stackrel{\text{def}}{=} \frac{\varpi_{rK}}{TOL_r}, \quad z_{\theta K} \stackrel{\text{def}}{=} \frac{\varpi_{\theta K}}{TOL_\theta};$

(c) $\Phi_K \stackrel{\text{def}}{=} \min(\phi_{rK}, \phi_{\theta K})$ where $\phi_{rK} \stackrel{\text{def}}{=} \left(\frac{\left(\frac{TOL_r}{\varpi_{r0}}\right)^{\frac{1}{pK_d}}}{\left(\frac{\varpi_{rK}}{\varpi_{r0}}\right)^{\frac{1}{pK}}}\right),$

$$\phi_{\theta K} \stackrel{\text{def}}{=} \left(\frac{\left(\frac{TOL_\theta}{\varpi_{\theta 0}}\right)^{\frac{1}{pK_d}}}{\left(\frac{\varpi_{\theta K}}{\varpi_{\theta 0}}\right)^{\frac{1}{pK}}}\right);$$

(5) IF TOLERANCE MET ($Z_K \leq 1$) AND $K < K_d$, THEN

(a) INCREMENT TIME: $t = t + \Delta t$;

(b) CONSTRUCT NEW TIME STEP: $\Delta t = \Phi_K \Delta t$;

(c) SELECT MINIMUM: $\Delta t = \min(\Delta t^{lim}, \Delta t)$ AND GO TO (1);

(6) IF TOLERANCE NOT MET ($Z_K > 1$) AND $K = K_d$, THEN:

(a) CONSTRUCT NEW TIME STEP: $\Delta t = \Phi_K \Delta t$;

(b) RESTART AT TIME $= t$ AND GO TO (1).

ALGORITHM 7.1

7.5.2 Semi-analytical examples

For the class of coupled systems considered in this work, the coupled operator's spectral radius is directly dependent on the time step discretization Δt. We consider a simple example that illustrates the essential concepts. Consider the coupling of two first-order equations and one second-order equation

$$\boxed{\begin{aligned} a\dot{w}_1 + w_2 &= 0,\\ b\dot{w}_2 + w_3 &= 0,\\ c\ddot{w}_3 + w_1 &= 0. \end{aligned}} \tag{7.46}$$

When this is discretized in time, for example, with a backward Euler scheme, we obtain

$$\boxed{\begin{aligned} \dot{w}_1{}^{L+1} &= \frac{w_1^{L+1} - w_1^L}{\Delta t},\\ \dot{w}_2{}^{L+1} &= \frac{w_2^{L+1} - w_2^L}{\Delta t},\\ \ddot{w}_3{}^{L+1} &= \frac{w_3^{L+1} - 2w_3^L + w_3^{L-1}}{(\Delta t)^2}, \end{aligned}} \tag{7.47}$$

which leads to the following coupled system:

$$\begin{bmatrix} 1 & \frac{\Delta t}{a} & 0 \\ 0 & 1 & \frac{\Delta t}{b} \\ \frac{(\Delta t)^2}{c} & 0 & 1 \end{bmatrix} \begin{Bmatrix} w_1^{L+1} \\ w_2^{L+1} \\ w_3^{L+1} \end{Bmatrix} = \begin{Bmatrix} w_1^L \\ w_2^L \\ 2w_3^L - w_3^{L-1} \end{Bmatrix}. \tag{7.48}$$

For a recursive staggering scheme of Jacobi type, where the updates are made only after one complete iteration, considered here only for algebraic simplicity, we have[37]

$$\begin{bmatrix} 1 & 0 & 0 \\ 0 & 1 & 0 \\ 0 & 0 & 1 \end{bmatrix} \begin{Bmatrix} w_1^{L+1,K} \\ w_2^{L+1,K} \\ w_3^{L+1,K} \end{Bmatrix} = \begin{Bmatrix} w_1^L \\ w_2^L \\ 2w_3^L - w_3^{L-1} \end{Bmatrix} - \begin{Bmatrix} \frac{\Delta t}{a} w_1^{L+1,K-1} \\ \frac{\Delta t}{b} w_2^{L+1,K-1} \\ \frac{(\Delta t)^2}{c} w_3^{L+1,K-1} \end{Bmatrix}. \tag{7.49}$$

Rewriting this in terms of the standard fixed-point form, $\mathcal{G}(\boldsymbol{w}^{L+1,K-1}) + \mathcal{R} = \boldsymbol{w}^{L+1,K}$, yields

$$\underbrace{\begin{bmatrix} 0 & \frac{\Delta t}{a} & 0 \\ 0 & 0 & \frac{\Delta t}{b} \\ \frac{(\Delta t)^2}{c} & 0 & 0 \end{bmatrix}}_{\mathcal{G}} \underbrace{\begin{Bmatrix} w_1^{L+1,K-1} \\ w_2^{L+1,K-1} \\ w_3^{L+1,K-1} \end{Bmatrix}}_{\boldsymbol{w}^{L+1,K-1}} + \underbrace{\begin{Bmatrix} w_1^L \\ w_2^L \\ 2w_3^L - w_3^{L-1} \end{Bmatrix}}_{\mathcal{R}} = \underbrace{\begin{Bmatrix} w_1^{L+1,K} \\ w_2^{L+1,K} \\ w_3^{L+1,K} \end{Bmatrix}}_{\boldsymbol{w}^{L+1,K}}. \tag{7.50}$$

[37]A Gauss–Seidel approach would involve using the most current iterate. Typically, under very general conditions, if the Jacobi method converges, the Gauss–Seidel method converges at a faster rate, while if the Jacobi method diverges, the Gauss–Seidel method diverges at a faster rate. For example, see Ames [5] for details. The Jacobi method is easier to address theoretically, so it is used for proof of convergence, and the Gauss–Seidel method is used at the implementation level.

The eigenvalues of $\mathcal{G}$ are given by $\lambda^3 = \frac{(\Delta t)^4}{abc}$ and, hence, for convergence we must have

$$|\max \lambda| = \left|\frac{(\Delta t)^4}{abc}\right|^{\frac{1}{3}} < 1. \tag{7.51}$$

We see that the spectral radius of the staggering operator grows quasi-linearly with the time step size, specifically superlinearly as $(\Delta t)^{\frac{4}{3}}$. Following Zohdi [208], a somewhat less algebraically complicated example illustrates a further characteristic of such solution processes. Consider the following example of reduced dimensionality, namely, a coupled first-order system

$$\boxed{\begin{aligned} a\dot{w}_1 + w_2 &= 0, \\ b\dot{w}_2 + w_1 &= 0. \end{aligned}} \tag{7.52}$$

When discretized in time with a backward Euler scheme and repeating the preceding procedure, we obtain the $\mathcal{G}$-form

$$\underbrace{\begin{bmatrix} 0 & \frac{\Delta t}{a} \\ \frac{\Delta t}{b} & 0 \end{bmatrix}}_{\mathcal{G}} \underbrace{\begin{Bmatrix} w_1^{L+1,K-1} \\ w_2^{L+1,K-1} \end{Bmatrix}}_{\boldsymbol{w}^{L+1,K-1}} + \underbrace{\begin{Bmatrix} w_1^{L} \\ w_2^{L} \end{Bmatrix}}_{\mathcal{R}} = \underbrace{\begin{Bmatrix} w_1^{L+1,K} \\ w_2^{L+1,K} \end{Bmatrix}}_{\boldsymbol{w}^{L+1,K}}. \tag{7.53}$$

The eigenvalues of $\mathcal{G}$ are

$$\lambda_{1,2} = \pm\sqrt{\frac{(\Delta t)^2}{ab}}. \tag{7.54}$$

We see that the convergence of the staggering scheme is directly related (linearly in this case) to the size of the time step. The solution to the example is

$$w_1^{L+1} = \frac{abw_1^L + b\Delta t w_2^L}{ab - (\Delta t)^2} = \underbrace{w_1^L - \frac{w_2^L}{a}\Delta t}_{\text{first staggered iteration}} + \underbrace{\frac{w_1^L}{ab}(\Delta t)^2}_{\text{second staggered iteration}} + \cdots \tag{7.55}$$

and

$$w_2^{L+1} = \frac{abw_2^L + a\Delta t w_1^L}{ab - (\Delta t)^2} = \underbrace{w_2^L - \frac{w_1^L}{b}\Delta t}_{\text{first staggered iteration}} + \underbrace{\frac{w_2^L}{ab}(\Delta t)^2}_{\text{second staggered iteration}} + \cdots. \tag{7.56}$$

As pointed out in Zohdi [208], the time step induced restriction for convergence matches the radius of analyticity of a Taylor series expansion of the solution around time t, which converges in a ball of radius from the point of expansion to the nearest singularity in Equations (7.55) and (7.56). In other words, the limiting step size is given by setting the denominator to zero,

$$ab - (\Delta t)^2 = 0, \tag{7.57}$$

which is in agreement with the condition derived from the analysis of the eigenvalues of $\mathcal{G}$.

Remark. Clearly, $1 \leq p \leq 2$ for a collection of first- and second-order equations. However, since we choose the individual field with the maximum error for time step adaptivity, we need to specifically use the corresponding convergence exponent (p) for the selected field's temporal discretization. If the equations of dynamic equilibrium of the particles are chosen, then $p = 2$, while if the equations of thermodynamic equilibrium of the particles are chosen, then $p = 1$. This issue is discussed further later in the analysis.

7.5.3 Numerical examples involving particulate flows

In order to simulate a particulate flow, we considered a group of N_p randomly positioned particles, of equal size, in a (starting) cubical domain of dimensions $D \times D \times D$, with D normalized to unity. The particle size and volume fraction were determined by a particle/sample size ratio, which was defined via a subvolume size $V \stackrel{\text{def}}{=} \frac{D\times D\times D}{N_p}$, where N_p was the number of particles in the entire cube. The ratio between the radius (b) and the subvolume was denoted by $\mathcal{L} \stackrel{\text{def}}{=} \frac{b}{V^{\frac{1}{3}}}$. The volume fraction occupied by the particles was $v_f \stackrel{\text{def}}{=} \frac{4\pi\mathcal{L}^3}{3}$. Thus, the total volume occupied by the particles, denoted by ν, could be written as $\nu = v_f N_p V$, and the total mass could be written as $M = \sum_{i=1}^{N_p} m_i = \rho\nu$, while that of an individual particle, assuming that all are the same size, was $m_i = \frac{\rho\nu}{N_p} = \rho\frac{4}{3}\pi b_i^3$. In order to visualize the flow clearly, we used $N_p = 100$ particles. The length scale of the particles was $\mathcal{L} = 0.25$, which resulted in a corresponding volume fraction of $v_f = \frac{4\pi\mathcal{L}^3}{3} = 0.0655$ and particulate radii of $b = 0.0539$. A mass density of the particles $= 2000$ kg/m^3 was used. The ambient temperature was selected to be $\theta_o = \theta_s = 300°$ K. The heat capacity of the particles was $C = 10^3$ J/kg °K, with emissivity of $\epsilon = 10^{-2}$. The critical temperature parameter in the coefficient of restitution relation was $\theta^* = 3000°$ K. The reaction constant was varied in the range 10^6 J/m$^2 \leq \kappa \leq 10^7$ J/m^2, with $I^* = 10^3 N$. The coefficient of convective heat transfer (h_c) was set to zero. We introduced the following near-field parameters per unit mass^2: $\alpha_{1ij} = \overline{\alpha}_1 m_i m_j$, $\alpha_{2ij} = \overline{\alpha}_2 m_i m_j$, and $\alpha_{aij} = \overline{\alpha}_a m_i m_j$. This allowed us to scale the strength of the interaction forces according to the mass of the particles.[38] The initial mean velocity field, componentwise, was $(1.0, 0.1, 0.1)$ m/s with initial random perturbations around the mean velocity of $(\pm 1.0, \pm 0.1, \pm 0.1)$ m/s, and a critical threshold velocity of $v^* = 10$ m/s in Equation (7.16). The simulation duration was set to 5 s, with an upper bound on the time step size of $\Delta t^{lim} = 10^{-2}$ s and a starting time step size of 10^{-3} s. The tolerances of both fields (TOL_r and TOL_θ) for the fixed-point iterations were set to 10^{-6} and the upper limit on the number of fixed-point iterations was set to $K^d = 10^2$.

Two main types of computational tests were conducted:

1. varying κ, for a given field strength, $\overline{\alpha}_1 = 0.5$ and $\overline{\alpha}_2 = 0.25$, with a clustering augmentation of $\overline{\alpha}_a = 1.75$ (forcing a small gap characterized by $d_a = 1.03(2b)$), $\beta_a = 1$, $\delta_a = 1.65(2b)$, and

2. varying κ, for a given field strength, $\overline{\alpha}_1 = 0.5$ and $\overline{\alpha}_2 = 0.25$, without a clustering augmentation.

[38] Although we did not consider particles of different sizes in this example, this decomposition allows us to easily take this into account. Also, we enforced the near-field stability condition by setting $(\beta_1, \beta_2) = (1, 2)$.

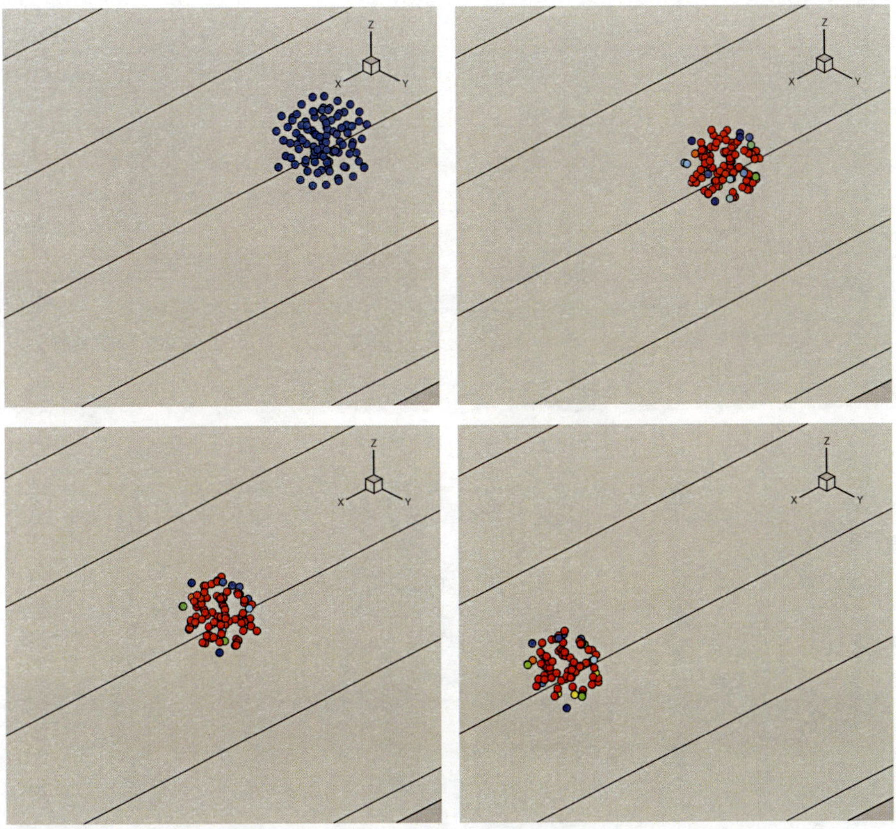

Figure 7.5. *Top to bottom and left to right, the dynamics of the particulate flow* with clustering forces: *An initially fine cloud of particles that clusters to form structures within the flow. Blue indicates a temperature of approximately* 300° K, *while red indicates a temperature of approximately* 400° K *(Zohdi* [217]*).*

For each different parameter selection, the initial conditions, i.e., random positions, velocities, temperatures, etc., were the same. We remark that parameter studies on the near-field strength, in isolation (without thermochemical coupling), have been conducted in Zohdi [209]. The field strength chosen was strong enough to induce vibratory motion and hence nonmonotone kinetic energy. Frames of the flows for cases 1 and 2, for (typical) values of $\kappa = 2 \times 10^6$ J/m^2, are shown in Figures 7.5 and 7.6. The plots in Figures 7.7–7.10 indicate the overall energetic and thermal behavior. Typically, the simulations took approximately between 1 min and 2 min on a standard (Dell, 2.33 GHz) laptop.[39] For the parameter ranges used in the presented simulations, the degree of adaptivity needed strongly depended on the presence of the clustering forces, and to a somewhat lesser degree on the thermochemical parameters. For example, for the 5-s simulation, if the time steps stayed at the starting value

[39]The computation time scales were, approximately, no worse than the number of particles squared. For example, a thousand particles took approximately 10 min.

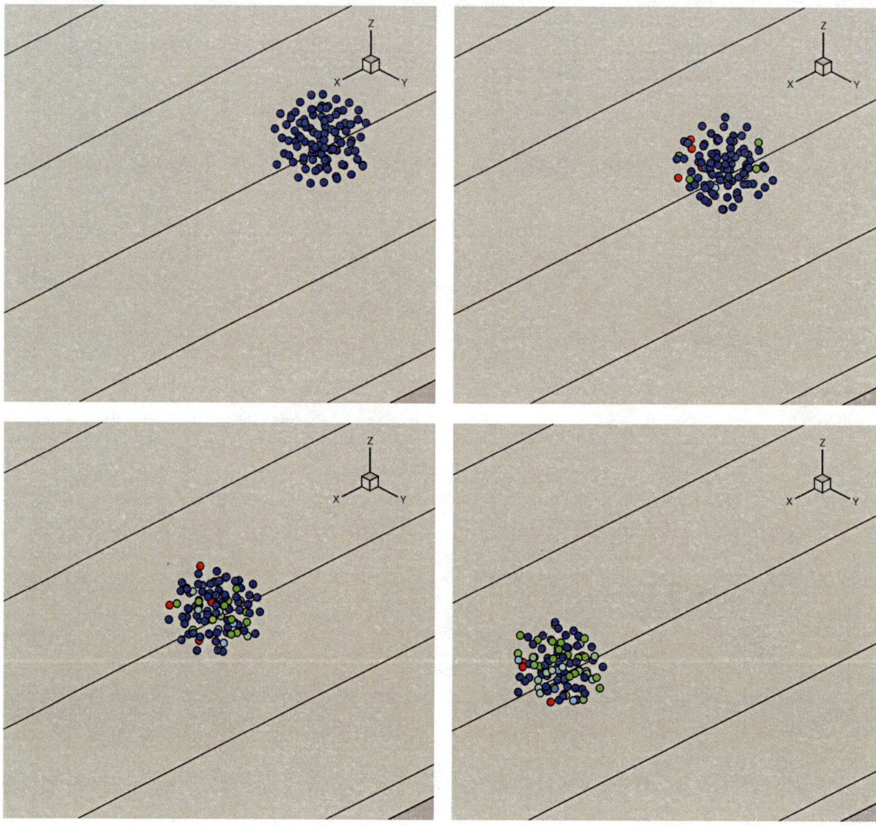

Figure 7.6. *Top to bottom and left to right, the dynamics of the particulate flow* without clustering forces. *Blue indicates a temperature of approximately* 300° K, *while red indicates a temperature of approximately* 400° K *(Zohdi* [217]*).*

($\Delta t = 10^{-3}$ s), then 5000 time steps would be needed if there had been no time step adaptivity (time step enlargement). Conversely, if the time steps were found to be unnecessarily small (an overkill) at the starting value ($\Delta t = 10^{-3}$ s), and, consequently, unrefined to the upper bound ($\Delta t^{lim} = 10^{-2}$ s), then approximately 500 time steps would be needed. Tables 7.1 and 7.2 indicate that, for the parameter ranges tested, when clustering forces were not present, the time steps did not need to be refined or unrefined. However, when clustering forces were present, the time steps could be unrefined for the given tolerances, requiring more internal fixed-point iterations. This was primarily because cluster structures formed, leading to fewer collisions between the larger objects, which did not require such small time steps (Figure 7.11). For the simulations with clustering forces, there was an expected thermal sensitivity. As the reaction constant κ became stronger, the number of fixed-point iterations required to achieve convergence increased. These results highlight an essential point of the adaptive time-stepping process, which is to allow the system to adjust to the physics of the problem. Some further remarks elaborating on this issue can be found in Zohdi [208]–[210].

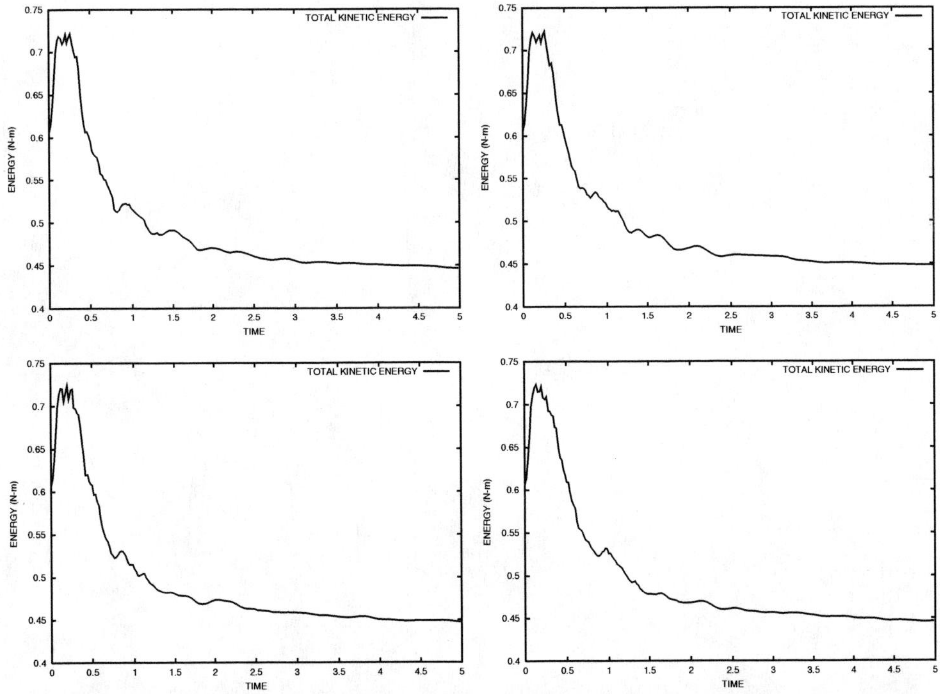

Figure 7.7. *Top to bottom and left to right,* with clustering forces: *the total kinetic energy in the system per unit mass with* $e_o = 0.5$, $\mu_s = 0.2$, $\mu_d = 0.1$, $\overline{\alpha}_1 = 0.5$, *and* $\overline{\alpha}_2 = 0.25$: (1) $\kappa = 10^6$ J/m^2, (2) $\kappa = 2 \times 10^6$ J/m^2, (3) $\kappa = 4 \times 10^6$ J/m^2, *and* (4) $\kappa = 8 \times 10^6$ J/m^2 *(Zohdi* [217]*).*

Qualitatively speaking, one should expect that, when a clustering field becomes active between two approaching particles, then kinetic energy is lost because of the disappearance of normal relative velocities between them. Conversely, kinetic energy is gained if the particles become dislodged, because the clustering field becomes inactive and the repulsive field suddenly dominates the remaining attractive forces, thus pushing the previously clustered particles away from one another. When the clustering binding field becomes active, the coefficient of restitution will play virtually no role, because the strength of the attractive force dominates everything. Thus, because the thermal field affects the particle dynamics through the coefficient of restitution, when clustering dominates, the particle dynamics will be only marginally affected by varying κ (Figure 7.7). However, the temperature of the particles in the presence of clustering will rise substantially, due to the large compressive forces between the contacting particles, which activate the chemical reactions. Also, we remark that the group dynamics, for different κ without clustering forces, deviate much more from one another than the cases when clustering is present (Figure 7.8). Typically, when two particles have clustered, since the binding field was strong, the particles rarely become dislodged. This issue has been been investigated in depth in Zohdi [225].

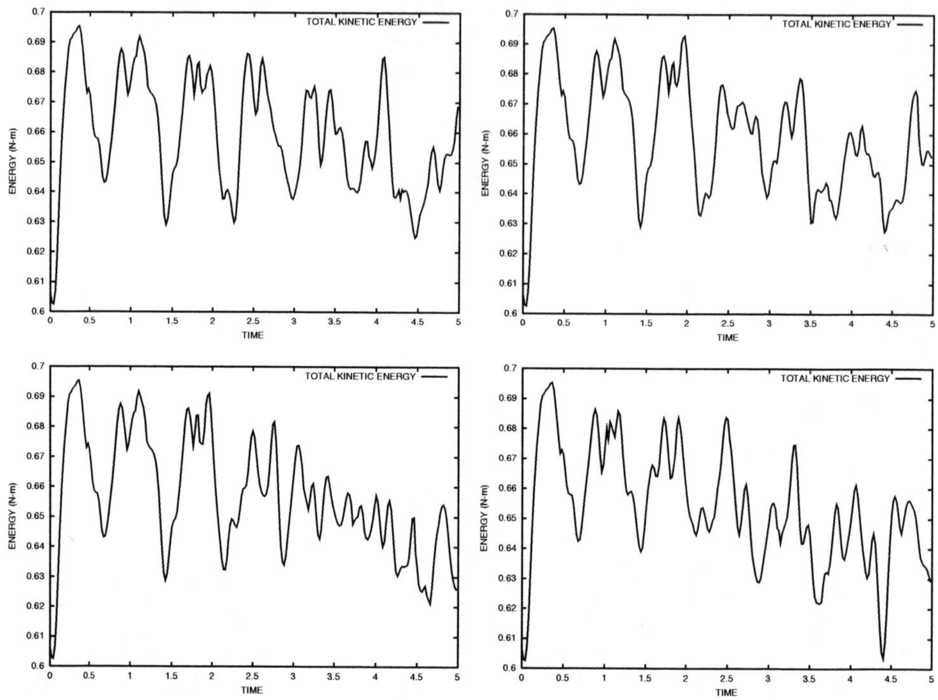

Figure 7.8. *Top to bottom and left to right,* without clustering forces: *the total kinetic energy in the system per unit mass with* $e_o = 0.5$, $\mu_s = 0.2$, $\mu_d = 0.1$, $\overline{\alpha}_1 = 0.5$, *and* $\overline{\alpha}_2 = 0.25$: (1) $\kappa = 10^6$ J/m^2, (2) $\kappa = 2 \times 10^6$ J/m^2, (3) $\kappa = 4 \times 10^6$ J/m^2, *and* (4) $\kappa = 8 \times 10^6$ J/m^2 *(Zohdi* [217]*).*

Remark. The interaction of clouds of granular gases with large (essentially immovable) obstacles arises in a variety of applications. It follows that associated impact phenomena are important. Accordingly, consider a stationary, massive obstacle ($M \gg m$) of radius b_{ob}. For this example, we assume that the obstacle has no near-field interaction with the particles, other than contact, which is governed by the classical expression for the ratio of the relative velocities before and after impact:

$$e \stackrel{\text{def}}{=} \frac{v_{obn}(t + \delta t) - v_{in}(t + \delta t)}{v_{in}(t) - v_{obn}(t)}, \tag{7.58}$$

where v_{obn} remains the same before and after impact. In Figure 7.13, the impact of a cloud against an obstacle is shown.[40] Let us focus on a particle impacting a massive obstacle $M \gg m$ (Figure 7.12). A balance of momentum reads for the particle as

$$mv(t) - \bar{I}\delta t \pm |\bar{F}|\delta t = mv(t + \delta t). \tag{7.59}$$

[40]All other parameters are the same as in the previous simulations.

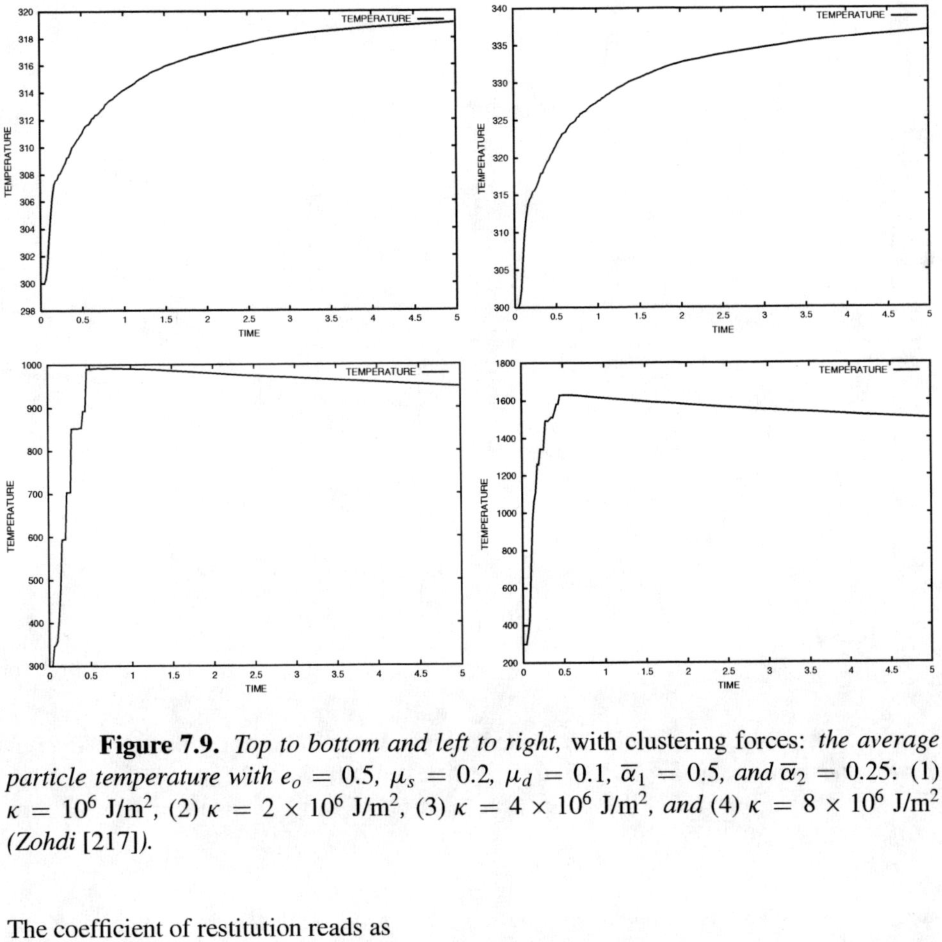

Figure 7.9. *Top to bottom and left to right,* with clustering forces: *the average particle temperature with* $e_o = 0.5$, $\mu_s = 0.2$, $\mu_d = 0.1$, $\overline{\alpha}_1 = 0.5$, *and* $\overline{\alpha}_2 = 0.25$: (1) $\kappa = 10^6$ J/m^2, (2) $\kappa = 2 \times 10^6$ J/m^2, (3) $\kappa = 4 \times 10^6$ J/m^2, *and* (4) $\kappa = 8 \times 10^6$ J/m^2 *(Zohdi* [217]*).*

The coefficient of restitution reads as

$$e \stackrel{\text{def}}{=} \frac{-v_{in}(t+\delta t)}{v_{in}(t)}, \tag{7.60}$$

so

$$\bar{I} = -\frac{m(v(t+\delta t) - v(t)}{\delta t} \pm |\bar{F}| = -\frac{mv(t)(1+e)}{\delta t} \pm |\bar{F}|, \tag{7.61}$$

where $\pm|\bar{F}|$ becomes $|\bar{F}|$ if attractive and $-|\bar{F}|$ if repulsive. Thus, we should expect that the impact of the aggregate will generally be lower if the interstitial forces are attractive at impact and that the impact of the aggregate will generally be higher if the interstitial forces are repulsive at impact. In order to illustrate this point, we consider two cases:

1. a given interaction field strength, $\overline{\alpha}_1 = 0.5$ and $\overline{\alpha}_2 = 0.25$,

2. no interaction field strength.

The results for a cloud of particles are shown in Figures 7.14 and 7.15.

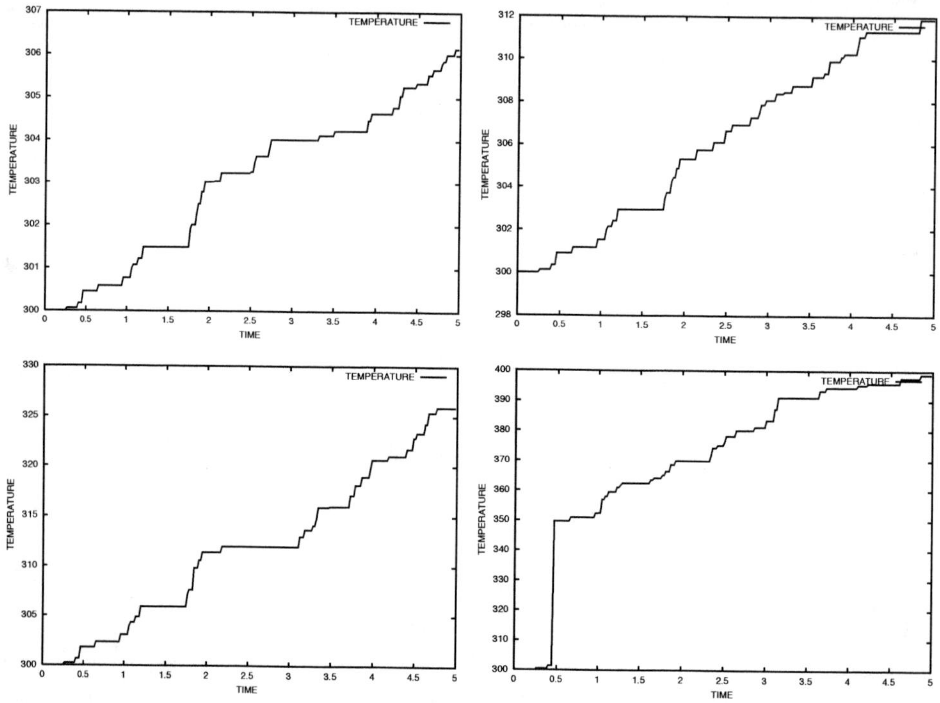

Figure 7.10. *Top to bottom and left to right,* without clustering forces: *the average particle temperature with* $e_o = 0.5$, $\mu_s = 0.2$, $\mu_d = 0.1$, $\overline{\alpha}_1 = 0.5$, *and* $\overline{\alpha}_2 = 0.25$: (1) $\kappa = 10^6$ J/m^2, (2) $\kappa = 2 \times 10^6$ J/m^2, (3) $\kappa = 4 \times 10^6$ J/m^2, *and* (4) $\kappa = 8 \times 10^6$ J/m^2 *(Zohdi* [217]*).*

Table 7.1. *The number of time steps and fixed-point iterations,* with clustering forces: *the average particle temperature with* $e_o = 0.5$, $\mu_s = 0.2$, $\mu_d = 0.1$, $\overline{\alpha}_1 = 0.5$, *and* $\overline{\alpha}_2 = 0.25$.

κ (J $\times$ 10^6/m^2)	Time Steps	Fixed-Point Iterations
1	586	1730
2	588	2076
4	598	4809
8	596	5584

Remark. Clearly, during flow processes, there is a possibility that the agglomerated clouds may impact one another and fragment as a result. In Figure 7.16, cloud collisions for slow approaching impact are shown, and in Figure 7.17 fast cloud impact is given.[41] A gallery of cloud interaction simulations can be found at `http://www.siam.org/books/cs04`.

[41] All other parameters are the same as in the previous simulations. In the case of slow impact, the clouds combine to form a larger cloud, and when the impact is fast, they disperse.

Table 7.2. *The number of time steps and fixed-point iterations,* without clustering forces: *the average particle temperature with* $e_o = 0.5$, $\mu_s = 0.2$, $\mu_d = 0.1$, $\overline{\alpha}_1 = 0.5$, *and* $\overline{\alpha}_2 = 0.25$.

κ ($\mathrm{J} \times 10^6/\mathrm{m}^2$)	Time Steps	Fixed-Point Iterations
1	5000	5025
2	5000	5024
4	5000	5029
8	5000	5024

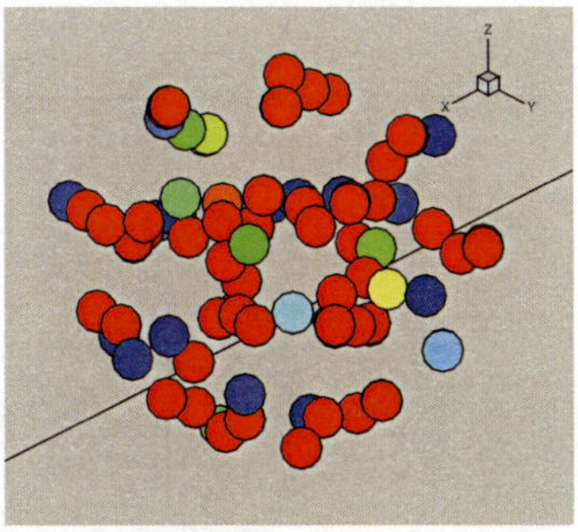

Figure 7.11. *A zoom on the structures that form* with clustering. *Blue indicates a temperature of approximately* 300° K, *while red indicates a temperature of approximately* 400° K *(Zohdi* [217]*).*

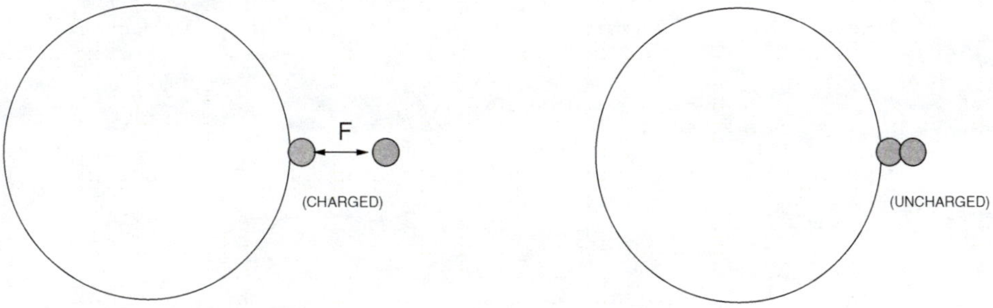

Figure 7.12. *Cases with and without charging.*

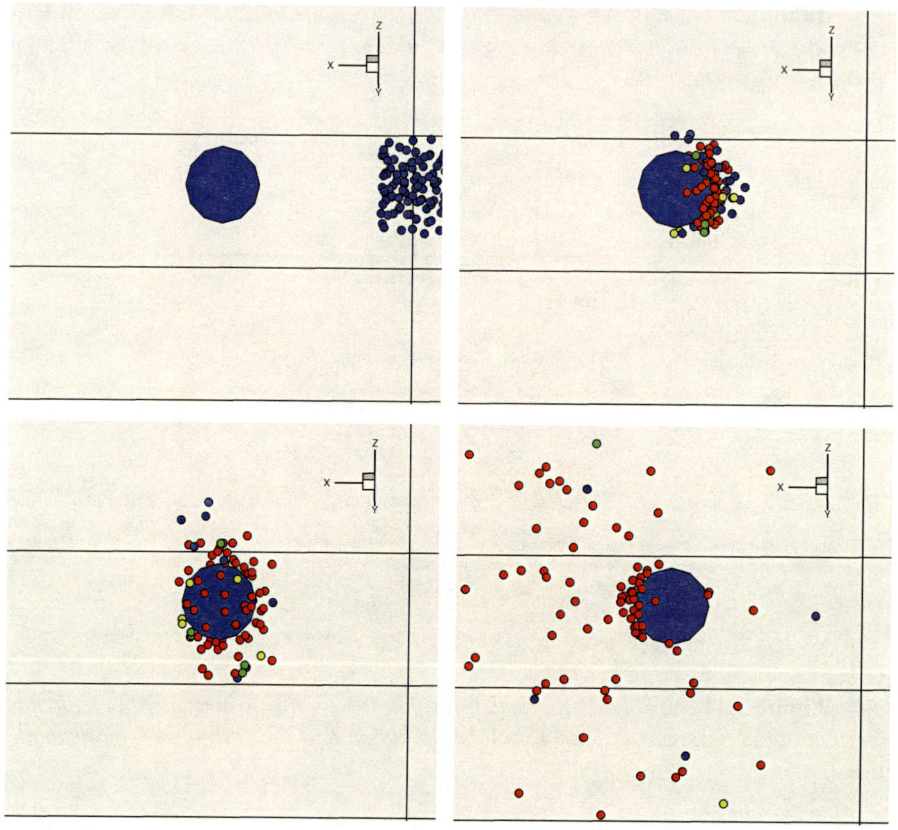

Figure 7.13. *Top to bottom and left to right, a charged cloud against an immovable obstacle.*

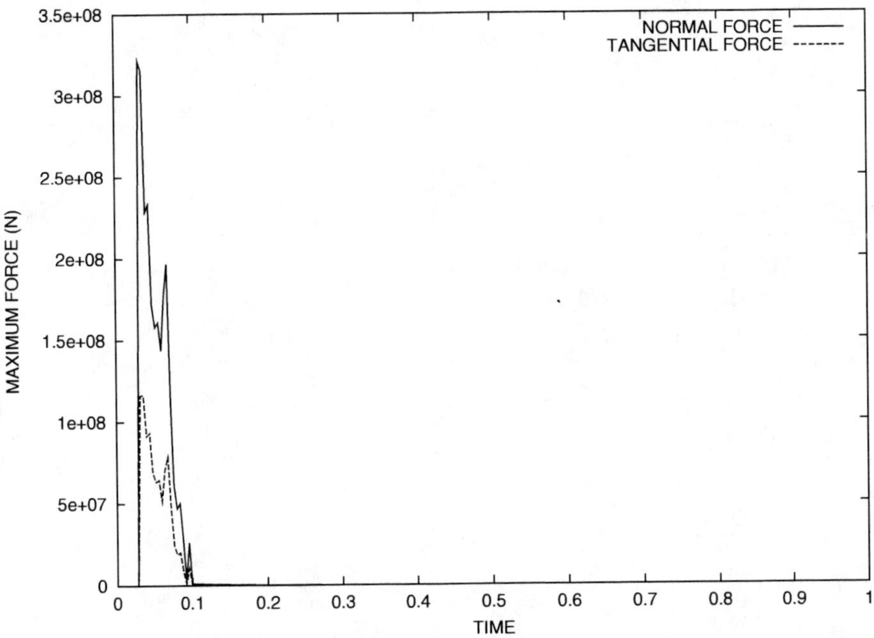

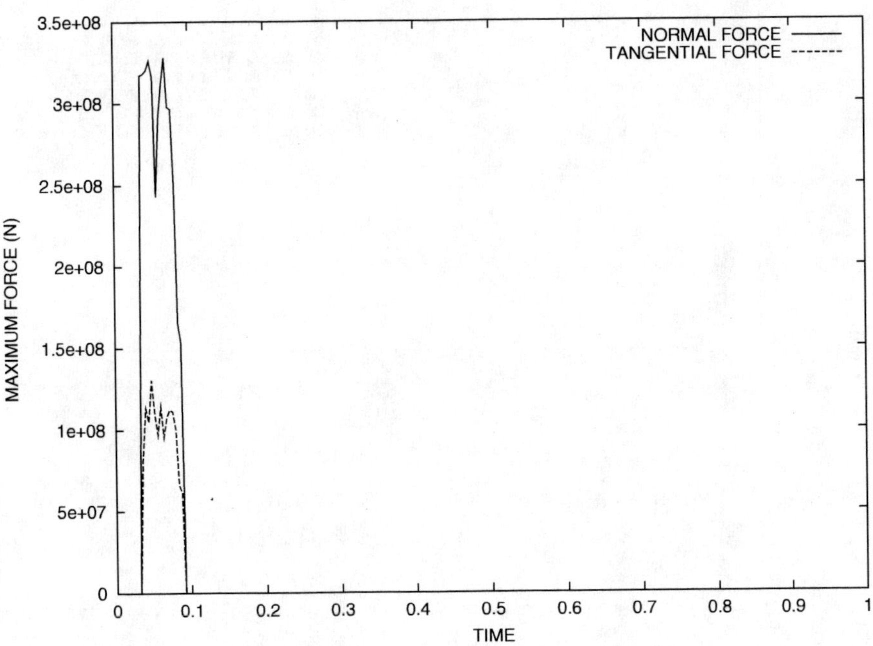

Figure 7.14. *The maximum force (and corresponding friction force) versus time imparted on the immovable obstacle surface,* $\max(\overline{I})$. *The top graph is with charging and the bottom is without charging. Notice that the maximum "signature" force is less with charging.*

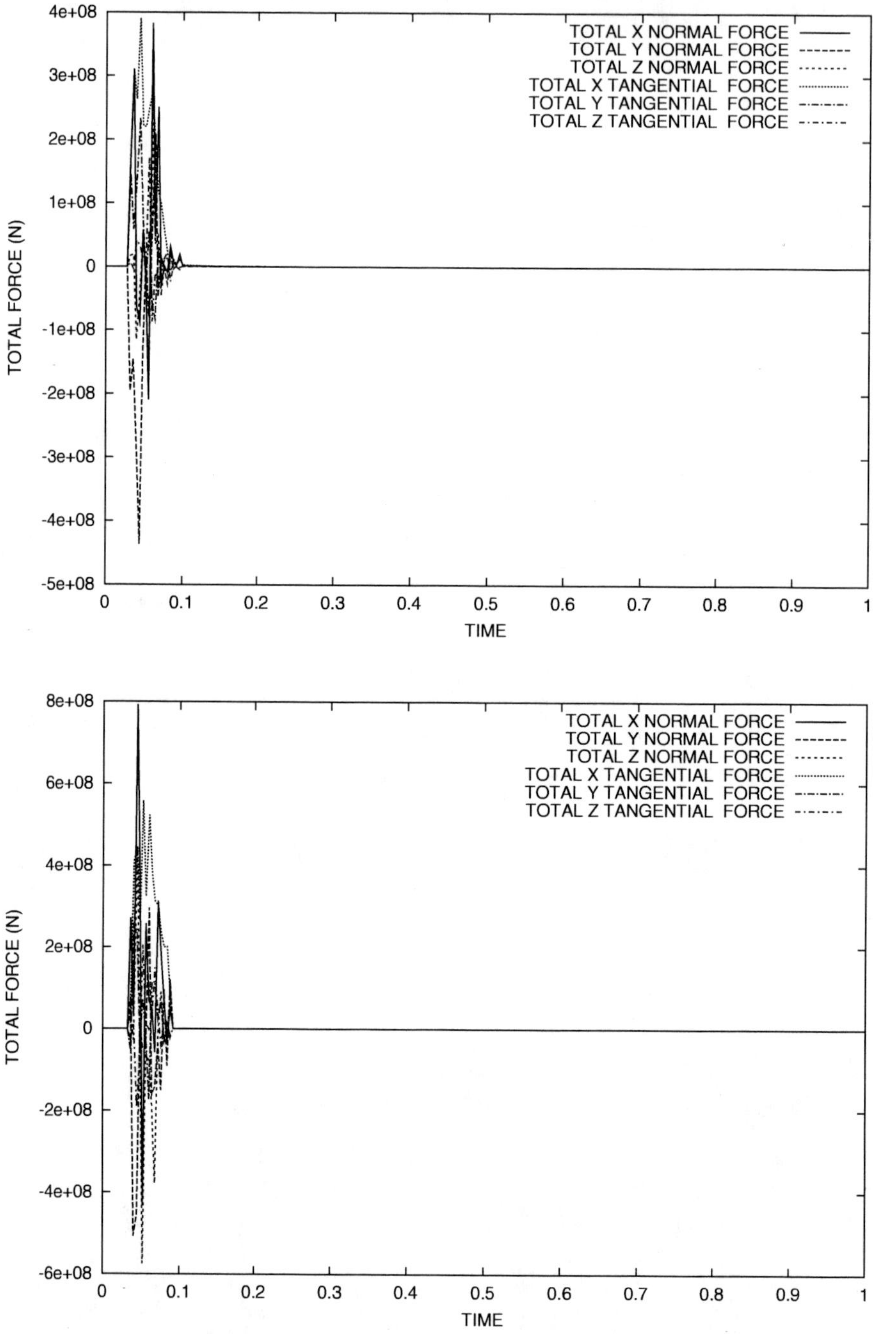

Figure 7.15. *The total force (and corresponding friction force) versus time imparted on the immovable obstacle surface,* $\max(\overline{I})$. *The top graph is with charging and the bottom is without charging. Notice that the total "signature" force is less with charging.*

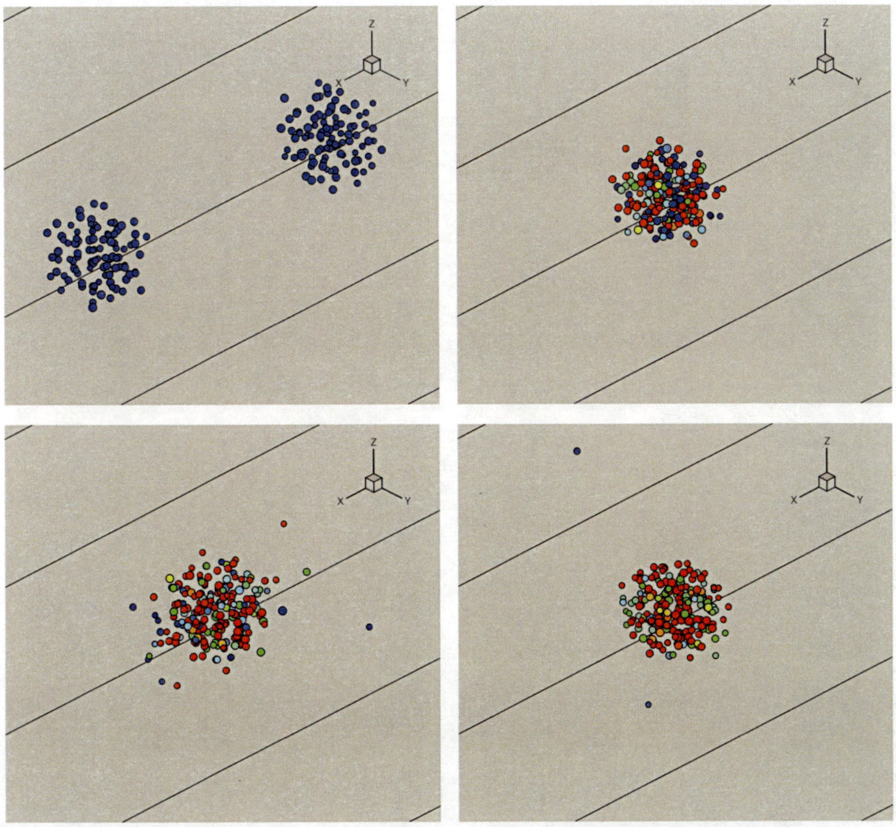

Figure 7.16. *Top to bottom and left to right,* slow impact *of charged clouds. The clouds combine into a larger cloud.*

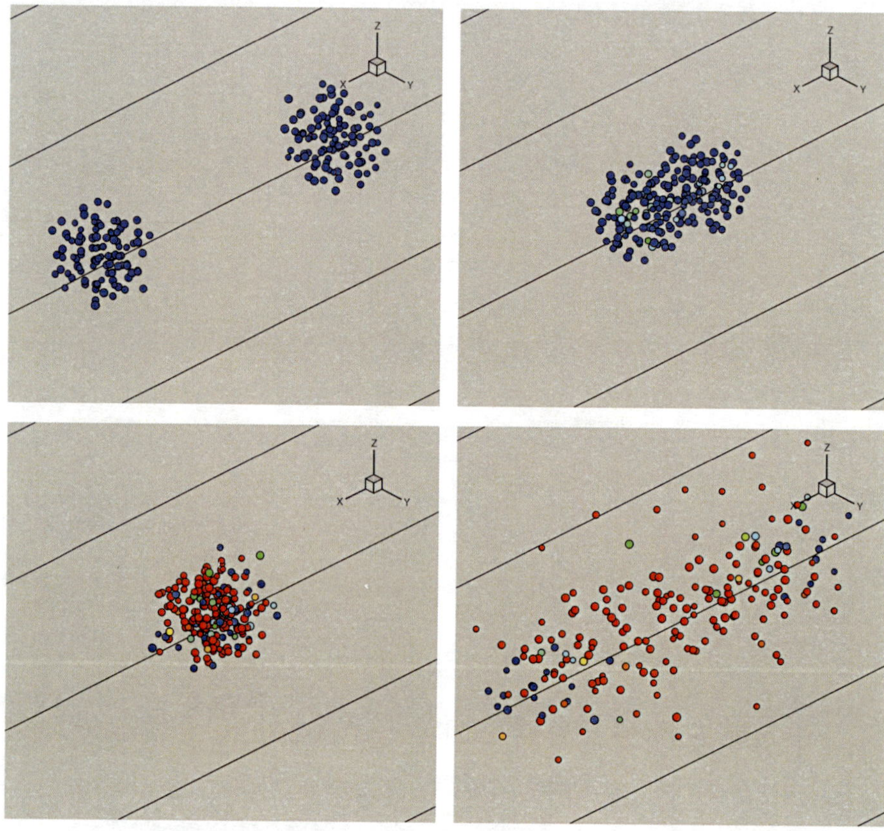

Figure 7.17. *Top to bottom and left to right,* fast impact *of charged clouds. The clouds disperse.*

Chapter 8

Coupled particle/fluid interaction

Until this point, we have ignored the presence of a fluid medium surrounding the particles. We now focus on the modeling and simulation of the dynamics of particles, coupled with a surrounding fluid, while bringing in several of the effects discussed earlier in the form of a model problem. Obviously, the number of research areas involving particles in a fluid undergoing various multifield processes is immense, and it would be futile to attempt to catalog all of the various applications. However, a common characteristic of such systems is that the various physical fields (thermal, mechanical, chemical, electrical, etc.) are strongly coupled.

This chapter develops a flexible and robust solution strategy to resolve coupled systems comprising large groups of flowing particles embedded within a fluid. A problem modeling groups of particles, which may undergo inelastic collisions in the presence of near-field forces, is considered. The particles are surrounded by a continuous interstitial fluid that is assumed to obey the compressible Navier–Stokes equations. Thermal effects are also considered. Such particle/fluid systems are strongly coupled due to the mechanical forces and heat transfer induced by the fluid on the particles and vice versa. Because the coupling of the various particle and fluid fields can dramatically change over the course of a flow process, a primary focus of this work is the development of a recursive "staggering" solution scheme, whereby the time steps are adaptively adjusted to control the error associated with the incomplete resolution of the coupled interaction between the various solid particulate and continuum fluid fields. A central feature of the approach is the ability to account for the presence of particles within the fluid in a straightforward manner that can be easily incorporated into any standard computational fluid mechanics code based on finite difference, finite element, or finite volume discretization. A three-dimensional example is provided to illustrate the overall approach.[42]

Remark. Although some portions of the presentation in this chapter may appear to be redundant with earlier parts of the monograph, there are subtle differences, and thus it is felt that a self-contained chapter is pedagogically superior to continual referral to previous portions of the monograph, which may lead to possible ambiguities.

---

[42] It is assumed that the particles are small enough that their rotation with respect to their mass centers is deemed insignificant. However, even in the event that the particles are not extremely small, we assume that any "spin" of the particles is small enough to neglect lift forces that may arise from the interaction with the surrounding fluid.

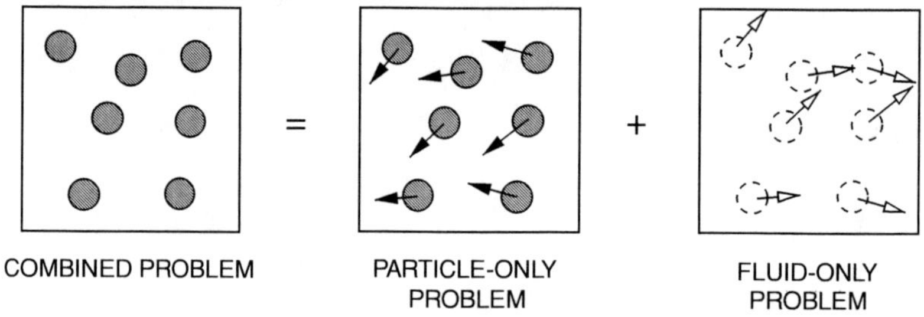

Figure 8.1. *Decomposition of the fluid/particle interaction (Zohdi [224]).*

8.1 A model problem

We consider a sufficiently complex model problem comprising of a group of nonintersecting spherical particles (N_p in total), each being small enough that their rotation with respect to their mass centers is deemed insignificant. The equation of motion for the ith particle in the system (Figure 8.1) is

$$m_i \ddot{\boldsymbol{r}}_i = \boldsymbol{\Psi}_i^{tot}(\boldsymbol{r}_1, \boldsymbol{r}_2, \ldots, \boldsymbol{r}_{N_p}), \tag{8.1}$$

where $\boldsymbol{r}_i$ is the position vector of the ith particle and $\boldsymbol{\Psi}_i^{tot}$ represents all forces acting on particle i. In particular, $\boldsymbol{\Psi}_i^{tot} = \boldsymbol{\Psi}_i^{drag} + \boldsymbol{\Psi}_i^{nf} + \boldsymbol{\Psi}_i^{con} + \boldsymbol{\Psi}_i^{fric}$ represents the forces due to fluid drag, near-field interaction, interparticle contact forces, and frictional forces. Clearly, under certain conditions one force may dominate over the others. However, this is generally impossible to ascertain a priori, since the dynamics and coupling in the system may change dramatically during the course of the flow process.

Remark. Throughout this chapter, boldface symbols indicate vectors or tensors. The inner product of two vectors $\boldsymbol{u}$ and $\boldsymbol{v}$ is denoted by $\boldsymbol{u} \cdot \boldsymbol{v}$. At the risk of oversimplification, we ignore the distinction between second-order tensors and matrices. Furthermore, we exclusively employ a Cartesian basis. Hence, if we consider the second-order tensor $\boldsymbol{A}$ with its matrix representation $[\boldsymbol{A}]$, then the product of two second-order tensors $\boldsymbol{A} \cdot \boldsymbol{B}$ is defined by the matrix product $[\boldsymbol{A}][\boldsymbol{B}]$, with components of $A_{ij}B_{jk} = C_{ik}$. The second-order inner product of two tensors or matrices is $\boldsymbol{A} : \boldsymbol{B} = A_{ij}B_{ij} = \mathrm{tr}([\boldsymbol{A}^T][\boldsymbol{B}])$. Finally, the divergence of a vector $\boldsymbol{u}$ is defined by $\nabla \cdot \boldsymbol{u} = u_{i,i}$, whereas for a second-order tensor $\boldsymbol{A}$, $\nabla \cdot \boldsymbol{A}$ describes a contraction to a vector with the components $A_{ij,j}$.

8.1.1 A simple characterization of particle/fluid interaction

We first consider drag force interactions between the fluid and the particles. The drag force acting on an object in a fluid flow (occupying domain ω and outward surface normal $\boldsymbol{n}$) is defined as

$$\boldsymbol{\Psi}^{drag} = \int_{\partial\omega} \boldsymbol{\sigma}_f \cdot \boldsymbol{n}\, dA, \tag{8.2}$$

where $\boldsymbol{\sigma}_f$ is the Cauchy stress. For a Newtonian fluid, $\boldsymbol{\sigma}_f$ is given by

$$\boldsymbol{\sigma}_f = -P_f\mathbf{1} + \lambda_f \mathrm{tr}\,\boldsymbol{D}_f\mathbf{1} + 2\mu_f\boldsymbol{D}_f = -P_f\mathbf{1} + 3\kappa_f\frac{\mathrm{tr}\,\boldsymbol{D}_f}{3}\mathbf{1} + 2\mu_f\boldsymbol{D}'_f, \tag{8.3}$$

where P_f is the thermodynamic pressure, $\kappa_f = \lambda_f + \frac{2}{3}\mu_f$ is the bulk viscosity, μ_f is the absolute viscosity, $\boldsymbol{D}_f = \frac{1}{2}(\nabla_x \boldsymbol{v}_f + (\nabla_x \boldsymbol{v}_f)^T)$ is the symmetric part of the velocity gradient, tr $\boldsymbol{D}_f$ is the trace of $\boldsymbol{D}_f$, and $\boldsymbol{D}'_f = \boldsymbol{D}_f - \frac{\text{tr}\,\boldsymbol{D}_f}{3}\mathbf{1}$ is the deviatoric part of $\boldsymbol{D}_f$. The stress is determined by solving the following coupled system of partial differential equations (compressible Navier–Stokes):

$$\text{Mass balance: } \frac{\partial \rho_f}{\partial t} = -\nabla_x \cdot (\rho_f \boldsymbol{v}_f),$$

$$\text{Energy balance: } \rho_f C_f \left(\frac{\partial \theta}{\partial t} + (\nabla_x \theta_f) \cdot \boldsymbol{v}_f \right) = \boldsymbol{\sigma}_f : \nabla_x \boldsymbol{v}_f + \nabla_x \cdot (\mathbb{K}_f \cdot \nabla \theta_f) + \rho_f z_f,$$

$$\text{Momentum balance: } \rho_f \left(\frac{\partial \boldsymbol{v}_f}{\partial t} + (\nabla_x \boldsymbol{v}_f) \cdot \boldsymbol{v}_f \right) = \nabla_x \cdot \boldsymbol{\sigma}_f + \rho_f \boldsymbol{b}_f, \tag{8.4}$$

where, at a point, ρ_f is the fluid density; $\boldsymbol{v}_f$ is the fluid velocity; θ_f is the fluid temperature; C_f is the fluid heat capacity; z_f is the heat source per unit mass; $\mathbb{K}_f$ is the thermal conductivity tensor, assumed to be isotropic of the form $\mathbb{K}_f = K_f \mathbf{1}$, K_f being the scalar thermal conductivity; and $\boldsymbol{b}_f$ represents body forces per unit mass. The thermodynamic pressure is given by an equation of state:

$$\mathcal{Z}(P_f, \rho_f, \theta_f) = 0. \tag{8.5}$$

The specific equation of state will be discussed later in the presentation.

The fluid domain will require spatial discretization with some type of mesh, for example, using a finite difference, finite volume, or finite element method. Usually, it is extremely difficult to resolve the flow in the immediate neighborhood of the particles, in particular if there are several particles. However, if the primary interest is in the dynamics of the particles, *as it is in this work*, an appropriate approach, which permits coarser discretization of the fluid continuum, is to employ effective drag coefficients, for example, defined via

$$C_D \stackrel{\text{def}}{=} \frac{||\boldsymbol{\Psi}_i^{drag}||}{\frac{1}{2}\langle \rho_f \rangle_{\omega_i} ||\langle \boldsymbol{v}_f \rangle_{\omega_i} - \boldsymbol{v}_i||^2 A_i}, \tag{8.6}$$

where $\langle (\cdot) \rangle_{\omega_i} \stackrel{\text{def}}{=} \frac{1}{|\omega_i|} \int_{\omega_i} (\cdot)\, d\omega_i$ is the volumetric average of the argument over the domain occupied by the ith particle, $\langle \boldsymbol{v}_f \rangle_{\omega_i}$ is the volumetric average of the fluid velocity, $\boldsymbol{v}_i$ is the velocity of the ith (solid) particle, and A_i is the cross-sectional area of the ith (solid) particle. For example, one possible way to represent the drag coefficient is with a piecewise definition, as a function of the Reynolds number (Chow [44]):

- For $0 < Re \leq 1$, $C_D = \frac{24}{Re}$.
- For $1 < Re \leq 400$, $C_D = \frac{24}{Re^{0.646}}$.
- For $400 < Re \leq 3 \times 10^5$, $C_D = 0.5$.
- For $3 \times 10^5 < Re \leq 2 \times 10^6$, $C_D = 0.000366 Re^{0.4275}$.
- For $2 \times 10^6 < Re < \infty$, $C_D = 0.18$.

Here, the local Reynolds number for a particle is $Re \stackrel{\text{def}}{=} \frac{2b_i \langle \rho_f \rangle_{\omega_i} ||\langle \boldsymbol{v}_f \rangle_{\omega_i} - \boldsymbol{v}_i||}{\mu}$ and b_i is the radius of the ith particle. The use of this simple concept makes it relatively straightforward to account for the presence of the solid particles in the fluid by augmenting the flow calculations with drag forces (Figure 8.1). Algorithmically speaking, one must compute the fluid flow with reaction forces due to the presence of the particles. To this end, one can use the volumetric forces ($\boldsymbol{b}_f$) and heat sources (z_f) within the fluid domain for this purpose by writing

$$
\boxed{
\begin{aligned}
&\rho_f\left(\frac{\partial \boldsymbol{v}_f}{\partial t} + (\nabla_{\boldsymbol{x}}\boldsymbol{v}_f)\cdot\boldsymbol{v}_f\right) = \nabla_{\boldsymbol{x}}\cdot\boldsymbol{\sigma}_f + \rho_f\boldsymbol{b}_f,\\
&\boldsymbol{b}_f = \boldsymbol{b}_D = -\frac{\boldsymbol{\Psi}_i^{drag}}{m_i} = -\frac{C_D\frac{1}{2}\langle\rho_f\rangle_{\omega_i}||\langle\boldsymbol{v}_f\rangle_{\omega_i}-\boldsymbol{v}_i||^2 A_i}{m_i}\boldsymbol{d} \qquad \left(\boldsymbol{d} = \frac{\langle\boldsymbol{v}_f\rangle_{\omega_i}-\boldsymbol{v}_i}{||\langle\boldsymbol{v}_f\rangle_{\omega_i}-\boldsymbol{v}_i||}\right),\\
&\rho_f C\left(\frac{\partial \theta_f}{\partial t} + (\nabla_{\boldsymbol{x}}\theta_f)\cdot\boldsymbol{v}_f\right) = \boldsymbol{\sigma}_f : \nabla_{\boldsymbol{x}}\boldsymbol{v}_f + \nabla_{\boldsymbol{x}}\cdot(\mathbb{K}_f\cdot\nabla_{\boldsymbol{x}}\theta_f) + \rho_f z_f,\\
&z_f = z_D = c_v|\boldsymbol{b}_D\cdot(\langle\boldsymbol{v}_f\rangle_{\omega_i} - \boldsymbol{v}_i)|,
\end{aligned}
}
\tag{8.7}
$$

where the drag force on the fluid, $\boldsymbol{b}_D$ (per unit mass), is nonzero if its location coincides with the particle domain and is zero otherwise. Here, z_D is the heat source due to the rate of work done by the drag force on the fluid.[43] Such source terms are easily projected onto a finite difference or finite element grid.[44] *This drag-based approach is designed to account for particles in the fluid using a coarse mesh. In other words, the smallest (mesh) scale allowable is that associated with the dimensions of the particles. Accordingly, we shall not employ meshes smaller than the particle length scale when simulations are performed later.*

Remark. More detailed analyses of fluid-particle interaction can be achieved in two primary ways: (1) direct, brute-force, numerical schemes, treating the particles as part of the fluid continuum (as another fluid or solid phase), and thus meshing them in a detailed manner, or (2) with semi-analytical techniques, such as those based on approximation of the interaction between the particles and the fluid, employing an analysis of the (interstitial) fluid gaps using lubrication theory. For a concise review of recent developments in such semi-analytical techniques, in particular methods that go beyond local analyses of flow in a single fluid gap, using discrete network approximations, which account for multiple hydrodynamic interactions, see Berlyand and Panchenko [30] and Berlyand et al. [31]. Although not employed here, discrete network approximations appear to be quite attractive for possibly improving the description of the interaction between the particles and the fluid, beyond a simple drag-based method (as adopted in this work), without resorting to detailed numerical meshing.

8.1.2 Particle thermodynamics

Throughout the thermal analysis of the particles, we shall use relatively simple models. Consistent with the particle-based philosophy, it is assumed that the temperature within each particle is uniform (a lumped mass approximation). We remark that the validity of assuming a uniform temperature within a particle is dictated by the Biot number. A small Biot number indicates that such an approximation is reasonable. The Biot number for a

[43] If the constant c_v is not selected as unity, this can indicate endothermic or exothermic particle/fluid chemical reactions.

[44] If the particles are significantly smaller than the mesh spacing, then the drag forces associated with the particles are computed from the nearest node/particle center pair.

sphere scales with the ratio of particle volume (V) to particle surface area (a_s), $\frac{V}{a_s} = \frac{b}{3}$, which indicates that a uniform temperature distribution is appropriate, since the particles are, by definition, small. Since it is assumed that the temperature fields are uniform within the particles, the gradient of the temperature within the particle is zero, i.e., $\nabla\theta = \mathbf{0}$. Therefore, a Fourier-type law for the heat flux will register a zero value, $\boldsymbol{q} = -\mathbb{K} \cdot \nabla\theta = \mathbf{0}$.

Under these assumptions, we consider an energy balance, governing the interconversions of mechanical, thermal, and chemical energy in a system, dictated by the first law of thermodynamics. Accordingly, we require the time rate of change of the sum of the kinetic energy ($\mathcal{K}$) and stored energy ($\mathcal{S}$) to be equal to the sum of the work rate (power, $\mathcal{P}$) and the net heat supplied ($\mathcal{H}$):

$$\frac{d}{dt}(\mathcal{K} + \mathcal{S}) = \mathcal{P} + \mathcal{H}, \tag{8.8}$$

where we assume that the stored energy is composed solely of a thermal part, $\mathcal{S} = mC\theta$, C being the heat capacity per unit mass. Consistent with the assumption that the particles deform negligibly during impact, *the amount of stored mechanical energy is deemed insignificant*. The kinetic energy is $\mathcal{K} = \frac{1}{2}m\boldsymbol{v} \cdot \boldsymbol{v}$. The mechanical power term is due to the forces acting on a particle:

$$\mathcal{P} = \frac{d\mathcal{W}}{dt} = \boldsymbol{\Psi}^{tot} \cdot \boldsymbol{v}. \tag{8.9}$$

For the particles, it is assumed that a process of convection, for example, governed by Newton's law of cooling and thermal radiation according to a simple Stefan–Boltzmann law, occurs. Accordingly, the first law reads

$$\underbrace{m\dot{\boldsymbol{v}} \cdot \boldsymbol{v} + mC\dot{\theta}}_{\frac{d(\mathcal{K}+\mathcal{S})}{dt}} = \underbrace{\boldsymbol{\Psi}^{tot} \cdot \boldsymbol{v}}_{power=\mathcal{P}} \underbrace{\underbrace{- h_c a_s(\theta - \theta_o)}_{convection} + \underbrace{mc_v|\boldsymbol{b}_D \cdot (\langle \boldsymbol{v}_f \rangle_\omega - \boldsymbol{v})|}_{drag} \underbrace{- \mathcal{B} a_s \epsilon(\theta^4 - \theta_s^4)}_{radiation}}_{\mathcal{H}}, \tag{8.10}$$

where θ_o is the temperature of the ambient fluid, h_c is the convection coefficient (using Newton's law of cooling), and θ_s is the temperature of the far-field surface (for example, a container surrounding the flow) with which radiative exchange is made. The Stefan–Boltzmann constant is $\mathcal{B} = 5.67 \times 10^{-8} \frac{W}{m^2-K}$; $0 \leq \epsilon \leq 1$ is the emissivity, which indicates how efficiently the surface radiates energy compared to a black-body (an ideal emitter); $0 \leq h_c$ is the convection coefficient (Newton's law of cooling); and a_s is the surface area of a particle. It is assumed that the radiation exchange between the particles is negligible.[45] Because $\frac{d\mathcal{K}}{dt} = m\dot{\boldsymbol{v}} \cdot \boldsymbol{v} = \boldsymbol{\Psi}^{tot} \cdot \boldsymbol{v} = \mathcal{P}$, we obtain a simplified form of the first law, $\frac{d\mathcal{S}}{dt} = \mathcal{H}$, and therefore Equation (8.10) becomes

$$mC\dot{\theta} = -h_c a_s(\theta - \theta_o) + mc_v|\boldsymbol{b}_D \cdot (\langle \boldsymbol{v}_f \rangle_\omega - \boldsymbol{v})| - \mathcal{B} a_s \epsilon(\theta^4 - \theta_s^4), \tag{8.11}$$

where $\theta_o = \langle \theta_f \rangle_\omega$ is the local average of the surrounding fluid temperature.

Remark. To account for the convective exchange between the fluid and the particles, we amend the source term in Equation (8.7) for the fluid to read

$$z_f = z_D = c_v|\boldsymbol{b}_f \cdot (\langle \boldsymbol{v}_f \rangle_\omega - \boldsymbol{v})| + \frac{h_c a_s(\theta - \theta_o)}{m}. \tag{8.12}$$

[45] Various arguments for such an assumption can be found in the classical text of Bohren and Huffman [33].

If the fluid is "radiationally" thick, then we assume that no radiation enters the system from the far field, namely, that $\mathcal{B}a_s\epsilon\theta_s^4 = 0$ in Equation (8.11), and that any emission from the particle gets absorbed by the fluid. Thus, in that situation, we can again amend the source term to read

$$z_f = z_D = c_v|\boldsymbol{b}_f \cdot (\langle \boldsymbol{v}_f\rangle_\omega - \boldsymbol{v})| + \frac{h_c a_s(\theta - \theta_o) + \mathcal{B}a_s\epsilon\theta^4}{m}. \tag{8.13}$$

Remark. We assume that various phenomena, such as near-field interaction, particle contact, interparticle friction, and particle thermal sensitivity, are similar for the wet and dry particulate flow problems, with the primary difference being that drag forces from the surrounding fluid need to be determined via numerical discretization of the Navier–Stokes equations, which is next.[46]

8.2 Numerical discretization of the Navier–Stokes equations

We now develop a fully implicit staggering scheme, in conjunction with a finite difference discretization, to solve the coupled system. Generally, such schemes proceed, within a discretized time step, by solving each field equation individually, allowing only the corresponding primary field variable (ρ_f, $\boldsymbol{v}_f$, or θ_f) to be active. This effectively (momentarily) decouples the system of differential equations. After the solution of each field equation, the primary field variable is updated, and the next field equation is solved in a similar manner, with only the corresponding primary variable being active. For accurate numerical solutions, the approach requires small time steps, primarily because the staggering error accumulates with each passing increment. Thus, such computations are usually computationally intensive.

First, let us consider a finite difference discretization of the derivatives in the governing equations where, for brevity, we write (L indicates the time step counter, $\boldsymbol{v}_f^L \stackrel{\text{def}}{=} \boldsymbol{v}_f(t)$, $\boldsymbol{v}_f^{L+1} \stackrel{\text{def}}{=} \boldsymbol{v}_f(t + \Delta t)$, etc.) for each finite difference node (i, j, k)

$$\boxed{\begin{aligned}
\rho_f^{i,j,k,L+1} &= \rho_f^{i,j,k,L} - \Delta t\left(\nabla_x \cdot (\rho_f \boldsymbol{v}_f)\right)^{i,j,k,L+1},\\
\mathcal{Z}(P_f^{i,j,k,L+1}&, \rho_f^{i,j,k,L+1}, \theta_f^{i,j,k,L+1}) = 0,\\
\theta_f^{i,j,k,L+1} &= \theta_f^{i,j,k,L} - \Delta t(\nabla_x\theta_f \cdot \boldsymbol{v}_f)^{i,j,k,L+1}\\
&\quad + \left(\frac{\Delta t}{\rho_f C_f}(\boldsymbol{\sigma}_f : \nabla_x \boldsymbol{v}_f + \nabla_x \cdot (\mathbb{K}_f \cdot \nabla_x \theta_f) + \rho_f z_f)\right)^{i,j,k,L+1},\\
\boldsymbol{v}_f^{i,j,k,L+1} &= \boldsymbol{v}_f^{i,j,k,L} - \Delta t(\nabla_x \boldsymbol{v}_f \cdot \boldsymbol{v}_f)^{i,j,k,L+1} + \frac{\Delta t}{\rho_f}\left(\nabla_x \cdot \boldsymbol{\sigma}_f + \rho_f \boldsymbol{b}_f\right)^{i,j,k,L+1},
\end{aligned}} \tag{8.14}$$

[46]Clearly, the wetting of the particle surfaces, breaking of hydrodymanic films, etc., are nontrivial, but are outside the scope of this introductory treatment.

where the derivatives are computed by the following:

$$
\begin{aligned}
\left(\frac{\partial \rho_f}{\partial t}\right)^{i,j,k,L} &\approx \frac{\rho_f(x_1,x_2,x_3,t+\Delta t)-\rho_f(x_1,x_2,x_3,t)}{\Delta t} \\
&= \frac{\rho_f^{i,j,k,L+1}-\rho_f^{i,j,k,L}}{\Delta t}, \\
\nabla_x \cdot (\rho_f \boldsymbol{v}_f) &\approx \frac{(\rho_f v_{f1})^{i+1,j,k,L}-(\rho_f v_{f1})^{i-1,j,k,L}}{2\Delta x_1} \\
&+ \frac{(\rho_f v_{f2})^{i,j+1,k,L}-(\rho_f v_{f2})^{i,j-1,k,L}}{2\Delta x_2} + \frac{(\rho_f v_{f3})^{i,j,k+1,L}-(\rho_f v_{f3})^{i,j,k-1,L}}{2\Delta x_3}
\end{aligned}
\tag{8.15}
$$

for the *continuity equation*;

$$
\begin{aligned}
&\left(\rho_f C_f \frac{\partial \theta_f}{\partial t}\right)^{i,j,k,L} \approx \rho_f^{i,j,k,L} C_f \frac{(\theta_f^{i,j,k,L+1}-\theta_f^{i,j,k,L})}{\Delta t}, \\
&(\rho_f C_f \nabla_x \theta_f \cdot \boldsymbol{v}_f)^{i,j,k,L} \approx \rho_f^{i,j,k,L} C_f \\
&\quad \times \left(v_{f1}^{i,j,k,L} \frac{\theta_f^{i+1,j,k,L}-\theta_f^{i-1,j,k,L}}{2\Delta x_1} + v_{f2}^{i,j,k,L} \frac{\theta_f^{i,j+1,k,L}-\theta_f^{i,j-1,k,L}}{2\Delta x_2} + v_{f3}^{i,j,k,L} \frac{\theta_f^{i,j,k+1,L}-\theta_f^{i,j,k-1,L}}{2\Delta x_3} \right), \\
&(\boldsymbol{\sigma}_f : \nabla_x \boldsymbol{v}_f)^{i,j,k,L} \approx \sigma_{f11}^{i,j,k,L} \frac{v_{f1}^{i+1,j,k,L}-v_{f1}^{i-1,j,k,L}}{2\Delta x_1} \\
&\quad + \sigma_{f22}^{i,j,k,L} \frac{v_{f2}^{i,j+1,k,L}-v_{f2}^{i,j-1,k,L}}{2\Delta x_2} + \sigma_{f33}^{i,j,k,L} \frac{v_{f3}^{i,j,k+1,L}-v_{f3}^{i,j,k-1,L}}{2\Delta x_3} \\
&\quad + \sigma_{f12}^{i,j,k,L} \left(\frac{v_{f1}^{i,j+1,k,L}-v_{f1}^{i,j-1,k,L}}{2\Delta x_2} + \frac{v_{f2}^{i+1,j,k,L}-v_{f2}^{i-1,j,k,L}}{2\Delta x_1} \right) \\
&\quad + \sigma_{f23}^{i,j,k,L} \left(\frac{v_{f2}^{i,j,k+1,L}-v_{f2}^{i,j,k-1,L}}{2\Delta x_3} + \frac{v_{f3}^{i,j+1,k,L}-v_{f3}^{i,j-1,k,L}}{2\Delta x_2} \right) \\
&\quad + \sigma_{f31}^{i,j,k,L} \left(\frac{v_{f3}^{i+1,J,k,L}-v_{f3}^{i-1,j,k,L}}{2\Delta x_1} + \frac{v_{f1}^{i,j,k+1,L}-v_{f1}^{i,j,k-1,L}}{2\Delta x_3} \right), \\
&(\nabla_x \cdot (\mathbb{K}_f \cdot \nabla_x \theta_f))^{i,j,k,L} \approx \mathbb{K}^{i,j,k,L} \frac{\theta_f^{i+1,j,k,L}-2\theta_f^{i,j,k,L}+\theta_f^{i-1,j,k,L}}{\Delta x_1^2} + \mathbb{K}_f^{i,j,k,L} \frac{\theta_f^{i,j+1,k,L}-2\theta_f^{i,j,k,L}+\theta_f^{i,j-1,k,L}}{\Delta x_2^2} \\
&\quad + \mathbb{K}_f^{i,j,k,L} \frac{\theta_f^{i,j,k+1,L}-2\theta_f^{i,j,k,L}+\theta_f^{i,j,k-1,L}}{\Delta x_3^2}
\end{aligned}
\tag{8.16}
$$

for the *balance of energy*; and

$$
\begin{aligned}
&\left(\frac{\partial v_{f1}}{\partial t}\right)^{i,j,k,L} \approx \frac{v_{f1}^{i,j,k,L+1} - v_{f1}^{i,j,k,L}}{\Delta t}, \quad \left(\frac{\partial v_{f2}}{\partial t}\right)^{i,j,k,L} \approx \frac{v_{f2}^{i,j,k,L+1} - v_{f2}^{i,j,k,L}}{\Delta t}, \\
&\left(\frac{\partial v_{f3}}{\partial t}\right)^{i,j,k,L} \approx \frac{v_{f3}^{i,j,k,L+1} - v_{f3}^{i,j,k,L}}{\Delta t}, \\
&((\nabla_x \boldsymbol{v}_f) \cdot \boldsymbol{v}_f)^{i,j,k,L} \approx v_{f1}^{i,j,k,L} \frac{v_{f1}^{i+1,j,k,L} - v_{f1}^{i-1,j,k,L}}{2\Delta x_1} \\
&\quad + v_{f2}^{i,j,k,L} \frac{v_{f1}^{i,j+1,k,L} - v_{f1}^{i,j-1,k,L}}{2\Delta x_2} + v_{f3}^{i,j,k,L} \frac{v_{f1}^{i,j,k+1,L} - v_{f1}^{i,j,k-1,L}}{2\Delta x_3} \\
&\quad + v_{f1}^{i,j,k,L} \frac{v_{f2}^{i+1,j,k,L} - v_{f2}^{i-1,j,k,L}}{2\Delta x_1} + v_{f2}^{i,j,k,L} \frac{v_{f2}^{i,j+1,k,L} - v_{f2}^{i,j-1,k,L}}{2\Delta x_2} \\
&\quad + v_{f3}^{i,j,k,L} \frac{v_{f2}^{i,j,k+1,L} - v_{f2}^{i,j,k-1,L}}{2\Delta x_3} + v_{f1}^{i,j,k,L} \frac{v_{f3}^{i+1,j,k,L} - v_{f3}^{i-1,j,k,L}}{2\Delta x_1} \\
&\quad + v_{f2}^{i,j,k,L} \frac{v_{f3}^{i,j+1,k,L} - v_{f3}^{i,j-1,k,L}}{2\Delta x_2} + v_{f3}^{i,j,k,L} \frac{v_{f3}^{i,j,k+1,L} - v_{f3}^{i,j,k-1,L}}{2\Delta x_3}, \\
&(\nabla_x \cdot \boldsymbol{\sigma}_f)^{i,j,k,L} \\
&\approx \left(\frac{\sigma_{f11}^{i+1,j,k,L} - \sigma_{f11}^{i-1,j,k,L}}{2\Delta x_1} + \frac{\sigma_{f12}^{i,j+1,k,L} - \sigma_{f12}^{i,j-1,k,L}}{2\Delta x_2} + \frac{\sigma_{f13}^{i,j,k+1,L} - \sigma_{f13}^{i,j,k-1,L}}{2\Delta x_3} \right) \boldsymbol{e}_1 \\
&+ \left(\frac{\sigma_{f21}^{i+1,j,k,L} - \sigma_{f21}^{i-1,j,k,L}}{2\Delta x_1} + \frac{\sigma_{f22}^{i,j+1,k,L} - \sigma_{f22}^{i,j-1,k,L}}{2\Delta x_2} + \frac{\sigma_{f23}^{i,j,k+1,L} - \sigma_{f23}^{i,j,k-1,L}}{2\Delta x_3} \right) \boldsymbol{e}_2 \\
&+ \left(\frac{\sigma_{f31}^{i+1,j,k,L} - \sigma_{f31}^{i-1,j,k,L}}{2\Delta x_1} + \frac{\sigma_{f32}^{i,j+1,k,L} - \sigma_{f32}^{i,j-1,k,L}}{2\Delta x_2} + \frac{\sigma_{f33}^{i,j,k+1,L} - \sigma_{f33}^{i,j,k-1,L}}{2\Delta x_3} \right) \boldsymbol{e}_3
\end{aligned}
\tag{8.17}
$$

for the *balance of linear monentum*. The discretized system is formulated next as an implicit time-stepping scheme within each time step L, whereby (1) one solves for the density, assuming the thermal and velocity fields fixed, (2) one solves for the temperature, assuming the density and velocity fields fixed, and then (3) one solves for the velocity, assuming the density and thermal fields fixed. Below, we formulate such a system, with an iterative counter K (within a time step), *for each finite difference node*:

$$
\boxed{
\begin{aligned}
&\rho_f^{i,j,k,L+1,K} = \rho_f^{i,j,k,L} - \Delta t \left(\nabla_x \cdot (\rho_f \boldsymbol{v}_f)_f\right)^{i,j,k,L+1,K-1}, \\
&\mathcal{Z}\left(P_f^{i,j,k,L+1,K-1}, \rho_f^{i,j,k,L+1,K}, \theta_f^{i,j,k,L+1,K-1}\right) = 0, \\
&\theta_f^{i,j,k,L,K} = \theta_f^{i,j,k,L} - \Delta t (\nabla_x \theta_f \cdot \boldsymbol{v}_f)^{i,j,k,L+1,K-1} \\
&\qquad + \left(\frac{\Delta t}{\rho_f C_f}\left(\boldsymbol{\sigma}_f : \nabla_x \boldsymbol{v}_f + \nabla_x \cdot (\mathbb{K}_f \cdot \nabla_x \theta_f) + \rho_f z_f\right)\right)^{i,j,k,L+1,K-1}, \\
&\boldsymbol{v}_f^{i,j,k,L+1,K} = \boldsymbol{v}_f^{i,j,k,L} - \Delta t (\nabla_x \boldsymbol{v}_f \cdot \boldsymbol{v}_f)^{i,j,k,L+1,K-1} + \left(\frac{\Delta t}{\rho_f}\left(\nabla_x \cdot \boldsymbol{\sigma}_f + \rho_f \boldsymbol{b}_f\right)\right)^{i,j,k,L+1,K-1}.
\end{aligned}
}
\tag{8.18}
$$

In an abstract setting, we have

$$
\boxed{
\begin{aligned}
&\mathcal{A}_{f1}\left(\underline{\rho_f^{L+1,K}}, \theta_f^{L+1,K-1}, \boldsymbol{v}_f^{L+1,K-1}\right) = \mathcal{F}_{f1}\left(\rho_f^{L+1,K-1}, \theta_f^{L+1,K-1}, \boldsymbol{v}_f^{L+1,K-1}, \ldots\right) \\
&\hfill \text{(CONTINUITY)}, \\
&\mathcal{A}_{f2}\left(\rho_f^{L+1,K}, \underline{\theta_f^{L+1,K}}, \boldsymbol{v}_f^{L+1,K-1}\right) = \mathcal{F}_{f2}\left(\rho_f^{L+1,K}, \theta_f^{L+1,K-1}, \boldsymbol{v}_f^{L+1,K-1}, \ldots\right) \\
&\hfill \text{(ENERGY)}, \\
&\mathcal{A}_{f3}\left(\rho_f^{L+1,K}, \theta_f^{L+1,K}, \underline{\boldsymbol{v}_f^{L+1,K}}\right) = \mathcal{F}_{f3}\left(\rho_f^{L+1,K}, \theta_f^{L+1,K}, \boldsymbol{v}_f^{L+1,K-1}, \ldots\right) \\
&\hfill \text{(MOMENTUM)},
\end{aligned}
}
\tag{8.19}
$$

where only the underlined variable is active (to be solved for) in the corresponding differential equation, and where K is an iteration counter.

8.3 Numerical discretization of the particle equations

As for the dry particulate cases, for the time discretization of the acceleration terms in the equations of motion (Equation (8.1)), *for each particle*, we write

$$
\ddot{\boldsymbol{r}}^{L+1} \approx \frac{\dot{\boldsymbol{r}}^{L+1} - \dot{\boldsymbol{r}}^{L}}{\Delta t} \approx \frac{\frac{\boldsymbol{r}^{L+1} - \boldsymbol{r}^{L}}{\Delta t} - \dot{\boldsymbol{r}}^{L}}{\Delta t} \approx \frac{\boldsymbol{r}^{L+1} - \boldsymbol{r}^{L}}{\Delta t^2} - \frac{\dot{\boldsymbol{r}}^{L}}{\Delta t}, \tag{8.20}
$$

which collapses to the familiar difference stencil of $\ddot{\boldsymbol{r}}^{L+1} \approx \frac{\boldsymbol{r}^{L+1} - 2\boldsymbol{r}^{L} + \boldsymbol{r}^{L-1}}{(\Delta t)^2}$ when the time step size is uniform. Inserting this into $m\ddot{\boldsymbol{r}} = \boldsymbol{\Psi}^{tot}(\boldsymbol{r})$ leads to

$$
\boldsymbol{r}^{L+1,K} \approx \frac{\Delta t^2}{m}\left(\boldsymbol{\Psi}^{tot}(\boldsymbol{r}^{L+1,K-1})\right) + \left(\boldsymbol{r}^{L} + \Delta t \dot{\boldsymbol{r}}^{L}\right). \tag{8.21}
$$

For the thermal behavior, after temporal integration with the previously used finite difference time step (for the fluid), we have from Equation (8.20)

$$\begin{aligned}\theta(t+\Delta t) = {} & \frac{mC}{mC+h_c a_s \Delta t}\theta(t) - \frac{\Delta t \mathcal{B} a_s \epsilon}{mC+h_c a_s \Delta t}\left(\theta^4(t+\Delta t)-\theta_s^4\right) \\ & + \frac{mc_v \Delta t |\boldsymbol{b}_D \cdot (\langle \boldsymbol{v}_f\rangle_\omega - \boldsymbol{v})|}{mC+h_c a_s \Delta t} + \frac{h_c a_s \Delta t \theta_o}{mC+h_c a_s \Delta t}.\end{aligned} \tag{8.22}$$

This implicit nonlinear equation for θ, *for each particle*, is recast as

$$\begin{aligned}\theta^{L+1,K} = {} & \frac{mC}{mC+h_c a_s \Delta t}\theta^L - \frac{\Delta t \mathcal{B} a_s \epsilon}{mC+h_c a_s \Delta t}\left((\theta^{L+1,K-1})^4-\theta_s^4\right) \\ & + \frac{mc_v \Delta t |\boldsymbol{b}_D \cdot (\langle \boldsymbol{v}_f^{L+1,K}\rangle_\omega - \boldsymbol{v}^{L+1,K})|}{mC+h_c a_s \Delta t} + \frac{h_c a_s \Delta t \theta_o}{mC+h_c a_s \Delta t}\end{aligned} \tag{8.23}$$

and is added into the fixed-point scheme with the equations of momentum balance and the equations governing the fluid mechanics. Concisely, the equations for the particle mechanics problem can be addressed in an abstract setting, whereby one solves for the particle positions, assuming the thermal fields fixed,

$$\mathcal{A}_{p1}(\underline{\boldsymbol{r}^{L+1,K}}, \theta^{L+1,K-1}) = \mathcal{F}_{p1}(\boldsymbol{r}^{L+1,K-1}, \theta^{L+1,K-1}), \tag{8.24}$$

and then one solves for the thermal fields, assuming the particle positions fixed,

$$\mathcal{A}_{p2}(\boldsymbol{r}^{L+1,K}, \underline{\theta^{L+1,K}}) = \mathcal{F}_{p2}(\boldsymbol{r}^{L+1,K}, \theta^{L+1,K-1}). \tag{8.25}$$

Both of these equations, and the equations for the fluid, are solved simultaneously with an *adaptive multifield* staggering scheme, which we shall discuss shortly.

Remark. In order to determine the thermal state of the particles when impact-induced reactions are significant, we shall decompose the heat generation and heat transfer processes into two stages. Stage I describes the extremely short time interval when impact occurs, $\delta t \ll \Delta t$, and accounts for the effects of chemical reactions, which are relevant in certain applications, and energy release due to mechanical straining. Stage II accounts for the postimpact behavior involving convective and radiative effects, as discussed earlier. As before, we consider an energy balance, governing the interconversions of mechanical, thermal, and chemical energy in a system, dictated by the first law of thermodynamics, $\frac{d}{dt}(\mathcal{K}+\mathcal{S}) = \mathcal{P}+\mathcal{H}$, with the previous assumptions leading to $\frac{d\mathcal{S}}{dt} = \mathcal{H}$. For Stage I, the primary source of heat is the chemical reactions that occur upon impact due to the presence of a reactive layer. The chemical reaction energy is defined as

$$\delta\mathcal{H} \stackrel{\text{def}}{=} \int_t^{t+\delta t} \mathcal{H}\, dt. \tag{8.26}$$

The first law can be rewritten for the temperature at time $= t+\delta t$ as

$$\theta(t+\delta t) = \theta(t) + \frac{\delta\mathcal{H}}{mC}. \tag{8.27}$$

The energy released from the reactions is assumed to be proportional to the amount of the fluid substance (for example, a gas) available to be compressed in the contact area between the particles. A typical, ad hoc approximation in combustion processes is to write, for example, a linear relation, with a saturation (limiting) value (ξ),

$$\delta\mathcal{H} \approx \xi \min\left(\frac{|\overline{I}_n|}{I_n^*}, 1\right)\pi b^2, \tag{8.28}$$

where ξ is the reaction constant (energy per unit area [J/m^2]), I_n^* is a normalization parameter, and b is the particle radius. For details on a variety of such relations, see, for example, Schmidt [172]. For the particle sizes and material properties of interest, the term $\frac{\delta\mathcal{H}}{mC}$ in Equation (8.27) indicates that

$$\delta\theta \stackrel{\text{def}}{=} \theta(t+\delta t) - \theta(t) = \frac{\delta\mathcal{H}}{mC} \propto \frac{\xi}{\rho C b}. \tag{8.29}$$

Thus, when values of ξ are chosen such that $\frac{\xi}{\rho C b} \gg 1$, this will generate a significant amount of heat.[47] Thereafter (Stage II, postimpact), it is assumed that a process of convection, for example, governed by Newton's law of cooling and radiation according to a simple Stefan–Boltzmann law, occurs. Since $\delta t \ll \Delta t$ we assign $\overline{\theta}^L = \theta(t+\delta t) = \theta(t) + \frac{\delta\mathcal{H}}{mC}$ and replace θ^L with it in Equation (8.23) to obtain

$$\begin{aligned}\theta(t+\Delta t) &= \frac{mC}{mC + h_c a_s \Delta t}\theta(t) - \frac{\Delta t \mathcal{B} a_s \epsilon}{mC + h_c a_s \Delta t}\left(\theta^4(t+\Delta t) - \theta_s^4\right) \\ &+ \frac{mc_v \Delta t |\boldsymbol{b}_D \cdot (\langle \boldsymbol{v}_f \rangle_\omega - \boldsymbol{v})|}{mC + h_c a_s \Delta t} + \frac{h_c a_s \Delta t \theta_o}{mC + h_c a_s \Delta t}.\end{aligned} \tag{8.30}$$

8.4 An adaptive staggering solution scheme

We now develop a temporally adaptive staggering scheme by extending the approach presented earlier. Let us denote the entire coupled system as $\mathcal{A}(\boldsymbol{w}^{L+1}) = \mathcal{F}$, where $\boldsymbol{w}$ is a multifield vector that represents the particle positions $(\boldsymbol{r})$, the particle temperatures (θ), the nodal fluid velocities $(\boldsymbol{v}_f)$, and the temperatures (θ_f), i.e., $\boldsymbol{w} = (\boldsymbol{r}, \theta, \boldsymbol{v}_f, \theta_f)$. It is convenient to write

$$\mathcal{A}(\boldsymbol{w}^{L+1}) - \mathcal{F} = \mathcal{G}(\boldsymbol{w}^{L+1}) - \boldsymbol{w}^{L+1} + \mathcal{R} = \boldsymbol{0}, \tag{8.31}$$

where $\mathcal{R}$ is a remainder term that does not depend on the solution, i.e., $\mathcal{R} \neq \mathcal{R}(\boldsymbol{w}^{L+1})$. A straightforward iterative scheme can be written as

$$\boldsymbol{w}^{L+1,K} = \mathcal{G}(\boldsymbol{w}^{L+1,K-1}) + \mathcal{R}, \tag{8.32}$$

where $K = 1, 2, 3, \ldots$ is the index of iteration within time step $L+1$. The convergence of such a scheme depends on the characteristics of $\mathcal{G}$. Namely, a sufficient condition for

[47] By construction, this model has increased heat production, via $\delta\mathcal{H}$, as κ increases.

convergence is that $\mathcal{G}$ is a contraction mapping for all $\boldsymbol{w}^{L+1,K}$, $K = 1, 2, 3 \dots$. In order to investigate this further, we define the staggering error as $\mathcal{E}^{L+1,K} = \boldsymbol{w}^{L+1,K} - \boldsymbol{w}^{L+1}$. A necessary restriction for convergence is iterative self-consistency, i.e., the "exact" (staggering error–free) solution must be represented by the scheme $\mathcal{G}(\boldsymbol{w}^{L+1}) + \mathcal{R} = \boldsymbol{w}^{L+1}$. Enforcing this restriction, a sufficient condition for convergence is the existence of a contraction mapping of the form

$$||\mathcal{E}^{L+1,K}|| \stackrel{\text{def}}{=} ||\boldsymbol{w}^{L+1,K} - \boldsymbol{w}^{L+1}|| = ||\mathcal{G}(\boldsymbol{w}^{L+1,K-1}) - \mathcal{G}(\boldsymbol{w}^{L+1})|| \leq \eta ||\boldsymbol{w}^{L+1,K-1} - \boldsymbol{w}^{L+1}||, \tag{8.33}$$

where, if $\eta < 1$ for each iteration K, then $\mathcal{E}^{L+1,K} \to \mathbf{0}$ for any arbitrary starting value $\boldsymbol{w}^{L+1,K=0}$ as $K \to \infty$. This contraction condition is sufficient, but not necessary, for convergence. For example, if we isolate the equation for the dynamics of the particles,

$$\boldsymbol{r}^{L+1,K} \approx \underbrace{\frac{\Delta t^2}{m}\left(\boldsymbol{\Psi}^{tot}(\boldsymbol{r}^{L+1,K-1})\right)}_{\mathcal{G}(\boldsymbol{r}^{L+1,K-1})} + \underbrace{\left(\boldsymbol{r}^L + \Delta t \dot{\boldsymbol{r}}^L\right)}_{\mathcal{R}}, \tag{8.34}$$

we observe that convergence is restricted by $\eta \propto EIG(\mathcal{G}) \propto \frac{\Delta t^2}{m}$. Thus, decreasing the time step size improves the convergence; however, we want to simultaneously maximize the time step sizes to decrease overall computing time while still meeting an error tolerance. In order to achieve this goal, we follow the approach provided earlier where (1) one approximates $\eta \approx S(\Delta t)^p$ (S is a constant) and (2) one assumes that the error within an iteration behaves approximately according to $(S(\Delta t)^p)^K ||\mathcal{E}^{L+1,0}|| = ||\mathcal{E}^{L+1,K}||$, $K = 1, 2, \dots$, where $||\mathcal{E}^{L+1,0}||$ is the initial norm of the iterative error and S is a function intrinsic to the system.[48] Our goal is to meet an error tolerance in exactly a preset number of iterations. To this end, we write this in the approximate form $(S(\Delta t_{\text{tol}})^p)^{K_d} ||\mathcal{E}^{L+1,0}|| = TOL$, where TOL is a tolerance and K_d is the desired number of iterations.[49] If the error tolerance is not met in the desired number of iterations, the contraction constant η is too large. Accordingly, we can solve for a new smaller step size, under the assumption that S is constant:[50]

$$\Delta t_{\text{tol}} = \Delta t \left(\frac{\left(\frac{TOL}{||\mathcal{E}^{L+1,0}||}\right)^{\frac{1}{pK_d}}}{\left(\frac{||\mathcal{E}^{L+1,K}||}{||\mathcal{E}^{L+1,0}||}\right)^{\frac{1}{pK}}} \right). \tag{8.35}$$

The assumption that S is constant is not critical, since the time steps are to be recursively refined and unrefined repeatedly. Clearly, the previous expression can also be used for time step enlargement if convergence is met in fewer than K_d iterations. Time step size adaptivity is paramount, since the flow's dynamics can dramatically change over the course of time, requiring radically different time step sizes for a preset level of accuracy. However, we must respect an upper bound dictated by the discretization error, i.e., $\Delta t \leq \Delta t^{lim}$. In order to couple this to the multifield computations, we define the normalized errors within each

[48]For the class of problems under consideration, due to the quadratic dependency on Δt, $p \approx 2$.

[49]Typically, K_d is chosen to be between five and ten iterations.

[50]In the definition of the error, since the "true" solution at a time step, $\boldsymbol{w}^{L+1}$, is unknown, we use the most current value of the solution, $\boldsymbol{w}^{L+1,K}$; thus, the error is to be interpreted as the relative error.

time step:

$$\mathcal{E}_{rK} \stackrel{\text{def}}{=} \frac{||\boldsymbol{r}^{L+1,K} - \boldsymbol{r}^{L+1,K-1}||}{||\boldsymbol{r}^{L+1,K} - \boldsymbol{r}^{L}||} \quad \text{and} \quad \mathcal{E}_{\theta K} \stackrel{\text{def}}{=} \frac{||\theta^{L+1,K} - \theta^{L+1,K-1}||}{||\theta^{L+1,K} - \theta^{L}||} \tag{8.36}$$

for the particles (summing over all particles) and

$$\hat{\mathcal{E}}_{r_f K} \stackrel{\text{def}}{=} \frac{||\boldsymbol{v}_f^{L+1,K} - \boldsymbol{v}_f^{L+1,K-1}||}{||\boldsymbol{v}_f^{L+1,K} - \boldsymbol{v}_f^{L}||} \quad \text{and} \quad \hat{\mathcal{E}}_{\theta_f K} \stackrel{\text{def}}{=} \frac{||\theta_f^{L+1,K} - \theta_f^{L+1,K-1}||}{||\theta_f^{L+1,K} - \theta_f^{L}||} \tag{8.37}$$

for the fluid (summing over all of the finite difference nodes). One can interpret these error metrics as the ratio of the staggering error to the change in the actual solution (from time step to time step). We now combine all of these (normalized) error metrics (ratios) into one single measure:

$$\mathcal{E}_{tot,K} = \frac{w_1\mathcal{E}_{rK} + w_2\mathcal{E}_{\theta K} + w_3\hat{\mathcal{E}}_{r_f K} + w_4\hat{\mathcal{E}}_{\theta_f K}}{w_1 + w_2 + w_3 + w_4}, \tag{8.38}$$

where the w_i's are weights. The overall algorithm is given as Algorithm 8.1. The purpose of the algorithm is to deliver solutions where the coupling is resolved in an iterative manner by the recursive staggered solution of the various field equations, constraints, etc. The incomplete coupling error is controlled by adaptively adjusting the time step sizes, while the temporal discretization accuracy dictates the upper limit on the time step size (Δt^{lim}).

Remark. As before, in Algorithm 8.1, at the implementation level, normalized (nondimensional) error measures were used. As with the unnormalized case, one approximates the error within an iteration to behave according to

$$(S(\Delta t)^p)^K \underbrace{\frac{||\boldsymbol{r}^{L+1,1} - \boldsymbol{r}^{L+1,0}||}{||\boldsymbol{r}^{L+1,0} - \boldsymbol{r}^{L}||}}_{\mathcal{E}_0} = \underbrace{\frac{||\boldsymbol{r}^{L+1,K} - \boldsymbol{r}^{L+1,K-1}||}{||\boldsymbol{r}^{L+1,K} - \boldsymbol{r}^{L}||}}_{\mathcal{E}_K}, \tag{8.39}$$

where the normalized measures characterize the ratio of the iterative (staggering) error within a time step to the difference in solutions between time steps. Since both $||\boldsymbol{r}^{L+1,0} - \boldsymbol{r}^{L}|| \approx \mathcal{O}(\Delta t)$ and $||\boldsymbol{r}^{L+1,K} - \boldsymbol{r}^{L}|| \approx \mathcal{O}(\Delta t)$ are of the same order, the use of normalized or unnormalized measures makes little difference in rates of convergence. However, the normalized measures are preferred since they have a clearer interpretation.

Remark. We remark that the forces needed to compute terms in the coefficient of restitution e, for example, E_{in}, E_{jn}, and D_{ij}, are obtained by using the most current known values of the $\boldsymbol{\Psi}_i$'s during the iterative solution process. In other words, the interaction forces are updated during the iterations, within a time step, based on the most current known positions of the particles. This process includes checking whether $||\boldsymbol{r}_i - \boldsymbol{r}_j|| \leq b_i + b_j$, which is a criterion for contact between particles.

Remark. For the fluid, notice that all of the contraction factors in Equation (8.19) scale as $\frac{\Delta t}{h}$ and $\frac{\Delta t}{h^2}$ (classical stability terms).

Remark. The result in Equation (7.51) provides a rough guide as to the selection of the exponent (p) for the overall system (when many different types of equations are present). The exponent p is approximately the sum of the product of each field equation that contains a numerical time derivative and the order of the corresponding time differentiation (first

(0) GLOBAL FIXED-POINT ITERATION (SET $i = 1$ AND $K = 0$):

(1) COMPUTE FLUID SOLUTION (FOR EACH NODE): $(\boldsymbol{v}_f, \rho_f, \theta_f)^{L+1,K}$ (FREEZING PARTICLE POSITIONS);

(2) IF $i > N_p$, THEN GO TO (4);

(3) IF $i \leq N_p$, THEN (FREEZING FLUID VARIABLES)

(a) COMPUTE POSITION: $\boldsymbol{r}_i^{L+1,K} = \frac{\Delta t^2}{m_i}\left(\boldsymbol{\Psi}_i^{tot}(\boldsymbol{r}^{L+1,K-1})\right) + \boldsymbol{r}_i^L + \Delta t \dot{\boldsymbol{r}}_i^L$;

(b) COMPUTE TEMPERATURE:

$$\theta_i^{L+1,K} = \frac{mC}{mC + h_c a_s \Delta t}\theta_i^{L+1,K-1} - \frac{\Delta t \mathcal{B} a_s \epsilon}{mC + h_c a_s \Delta t}\left((\theta_i^{L+1,K-1})^4 - \theta_s^4\right)$$
$$+\frac{h_c a_s \Delta t \theta_o}{mC + h_c a_s \Delta t} + \frac{m c_v \Delta t |\boldsymbol{b}_D \cdot (\langle \boldsymbol{v}_f^{L+1,K}\rangle_{\omega_i} - \boldsymbol{v}^{L+1,K})|}{mC + h_c a_s \Delta t};$$

(c) GO TO (2) AND NEXT FLOW PARTICLE ($i = i + 1$);

(4) ERROR MEASURES:

(a) $\mathcal{E}_{rK} \stackrel{\text{def}}{=} \frac{\sum_{i=1}^{N_p} ||\boldsymbol{r}_i^{L+1,K} - \boldsymbol{r}_i^{L+1,K-1}||}{\sum_{i=1}^{N_p} ||\boldsymbol{r}_i^{L+1,K} - \boldsymbol{r}_i^L||}$, $\quad \mathcal{E}_{\theta K} \stackrel{\text{def}}{=} \frac{\sum_{i=1}^{N_p} ||\theta_i^{L+1,K} - \theta_i^{L+1,K-1}||}{\sum_{i=1}^{N_p} ||\theta_i^{L+1,K} - \theta_i^L||}$,

$$\hat{\mathcal{E}}_{r_f K} \stackrel{\text{def}}{=} \frac{||\boldsymbol{v}_f^{L+1,K} - \boldsymbol{v}_f^{L+1,K-1}||}{||\boldsymbol{v}_f^{L+1,K} - \boldsymbol{v}_f^L||}, \quad \hat{\mathcal{E}}_{\theta_f K} \stackrel{\text{def}}{=} \frac{||\theta_f^{L+1,K} - \theta_f^{L+1,K-1}||}{||\theta_f^{L+1,K} - \theta_f^L||};$$

(b) $\mathcal{E}_{tot,K} = \frac{w_1 \mathcal{E}_{rK} + w_2 \mathcal{E}_{\theta K} + w_3 \hat{\mathcal{E}}_{rK} + w_4 \hat{\mathcal{E}}_{\theta K}}{w_1 + w_2 + w_3 + w_4}$,

$$TOL_{tot} = \frac{w_1 TOL_r + w_2 TOL_\theta + w_3 TOL_{r_f} + w_4 TOL_{\theta_f}}{w_1 + w_2 + w_3 + w_4};$$

(c) $\Phi_K \stackrel{\text{def}}{=} \left(\frac{\left(\frac{TOL_{tot}}{\mathcal{E}_{tot,0}}\right)^{\frac{1}{pK_d}}}{\left(\frac{\mathcal{E}_{tot,K}}{\mathcal{E}_{tot,0}}\right)^{\frac{1}{pK}}} \right)$;

(5) IF TOLERANCE MET ($\mathcal{E}_{tot,K} \leq 1$) AND $K < K_d$, THEN

(a) INCREMENT TIME: $t = t + \Delta t$;

(b) CONSTRUCT NEW TIME STEP: $\Delta t = \Phi_K \Delta t$;

(c) SELECT MINIMUM $\Delta t = \min(\Delta t^{lim}, \Delta t)$ AND GO TO (0);

(6) IF TOLERANCE NOT MET ($\mathcal{E}_{tot,K} > TOL$) AND $K = K_d$, THEN

(a) CONSTRUCT NEW TIME STEP: $\Delta t = \Phi_K \Delta t$;

(b) RESTART AT TIME $= t$ AND GO TO (0).

ALGORITHM 8.1

order, second order, etc.), divided by the sum of the number of equations using numerical time derivatives in the system. Explicitly,

$$p \approx \frac{\sum_{i=1}^{N} O_i}{N}, \tag{8.40}$$

where N is the number of field equations where a numerical derivative was used and O_i is the order of the time differentiation (first order, second order, etc.) of the individual field equation i.

Remark. An alternative and more severe way to measure the error is to define "violation ratios," i.e., the measure of which field is relatively more in error, compared to its corresponding tolerance, via $Z_K \stackrel{\text{def}}{=} \max(z_{rK}, z_{\theta K}, \hat{z}_{vK}, \hat{z}_{\theta_f K})$, where

$$z_{rK} \stackrel{\text{def}}{=} \frac{\mathcal{E}_{rK}}{TOL_r} \quad \text{and} \quad z_{\theta K} \stackrel{\text{def}}{=} \frac{\mathcal{E}_{\theta K}}{TOL_\theta} \tag{8.41}$$

and

$$\hat{z}_{vK} \stackrel{\text{def}}{=} \frac{\hat{\mathcal{E}}_{vK}}{TOL_v} \quad \text{and} \quad \hat{z}_{\theta_f K} \stackrel{\text{def}}{=} \frac{\hat{\mathcal{E}}_{\theta K}}{TOL_{\theta_f}}, \tag{8.42}$$

and then a minimum scaling factor $\Phi_K \stackrel{\text{def}}{=} \min(\phi_{rK}, \phi_{\theta K}, \hat{\phi}_{vK}, \hat{\phi}_{\theta_f K})$, where, for the particles

$$\phi_{rK} \stackrel{\text{def}}{=} \left(\frac{\left(\frac{TOL_r}{\mathcal{E}_{r0}}\right)^{\frac{1}{pK_d}}}{\left(\frac{\mathcal{E}_{rK}}{\mathcal{E}_{r0}}\right)^{\frac{1}{pK}}} \right), \quad \phi_{\theta K} \stackrel{\text{def}}{=} \left(\frac{\left(\frac{TOL_\theta}{\mathcal{E}_{\theta 0}}\right)^{\frac{1}{pK_d}}}{\left(\frac{\mathcal{E}_{\theta K}}{\mathcal{E}_{\theta 0}}\right)^{\frac{1}{pK}}} \right) \tag{8.43}$$

and for the fluid

$$\hat{\phi}_{vK} \stackrel{\text{def}}{=} \left(\frac{\left(\frac{\hat{TOL}_v}{\hat{\mathcal{E}}_{v0}}\right)^{\frac{1}{pK_d}}}{\left(\frac{\hat{\mathcal{E}}_{vK}}{\hat{\mathcal{E}}_{v0}}\right)^{\frac{1}{pK}}} \right), \quad \hat{\phi}_{\theta_f K} \stackrel{\text{def}}{=} \left(\frac{\left(\frac{\hat{TOL}_\theta}{\hat{\mathcal{E}}_{\theta 0}}\right)^{\frac{1}{pK_d}}}{\left(\frac{\hat{\mathcal{E}}_{\theta_f K}}{\hat{\mathcal{E}}_{\theta_f 0}}\right)^{\frac{1}{pK}}} \right). \tag{8.44}$$

However, in such an approach, if the individual field with the maximum error is used for time step adaptivity, we need to specifically use the corresponding convergence exponent (p) for the selected field's temporal discretization. If the equations of dynamic equilibrium of the particles are the field chosen, then $p = 2$. If the equations of thermodynamic equilibrium of the particles are the field chosen, then $p = 1$. If the equations of dynamic equilibrium of the fluid are the field chosen, then $p = 1$. If the equations of thermodynamic equilibrium of the fluid are the field chosen, then $p = 1$. However, this approach has some major drawbacks when many disparate fields are present. Specifically, when the maximum error measure oscillates from field to field within a time step or abruptly from time step to time step, convergence becomes quite difficult. Using the combined metric (Equation (8.38)) is more stable and, thus, preferred.

8.5 A numerical example

As a model problem, we considered a cubical representative volume of a particle-laden fluid flow (Figure 8.2). The classical random sequential addition algorithm was used to initially place nonoverlapping spherical particles into the domain of interest (Widom [200]). This

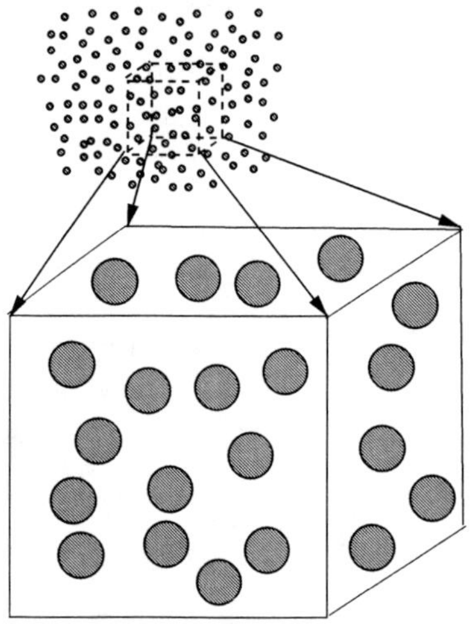

Figure 8.2. *A representative volume element extracted from a flow (Zohdi* [224]*).*

algorithm was adequate for the volume fraction ranges of interest (under 30%), since the limit of the method is on the order of 38%.

Any particles that exited a boundary were given the same velocity (now incoming) on the opposite boundary. Periodicity conditions were used to generate any numerical derivatives for finite difference stencils that extended beyond the boundary. Clearly, under these conditions the group velocity of the particles will tend toward the velocity of the ("background") fluid specified (controlled) on the boundary.

A Boussinesq-type (perturbation from an ideal gas) relation, adequate to describe dense gases, and fluids, was used for the equation of state, stemming from

$$\rho_f \approx \rho_o(\theta_o, P_o) + \left.\frac{\partial \rho_f}{\partial P_f}\right|_{\theta} \Delta P_f + \left.\frac{\partial \rho_f}{\partial \theta_f}\right|_{P_f} \Delta \theta_f, \tag{8.45}$$

where ρ_o, θ_o, P_o are reference values, $\Delta P_f = P_f - P_o$, and $\Delta\theta_f = \theta_f - \theta_o$. We define the thermal expansion as

$$\zeta_\theta \stackrel{\text{def}}{=} -\frac{1}{\rho_f}\left.\frac{\partial \rho_f}{\partial \theta_f}\right|_{P_f} = \frac{1}{V_f}\left.\frac{\partial V_f}{\partial \theta_f}\right|_{P_f} \tag{8.46}$$

and the bulk (compressibility) modulus by

$$\zeta_{com} \stackrel{\text{def}}{=} -V_f\left.\frac{\partial P_f}{\partial V_f}\right|_{\theta_f} = \rho_f\left.\frac{\partial P_f}{\partial \rho_f}\right|_{\theta_f}, \tag{8.47}$$

yielding the desired result

$$\rho_f \approx \rho_o\left(1 - \zeta_\theta \Delta\theta_f + \frac{1}{\zeta_{com}}\Delta P_f\right), \tag{8.48}$$

leading to

$$P_f \approx P_o + \zeta_{com}\left(\frac{\rho_f}{\rho_o} - 1 + \zeta_\theta \Delta\theta_f\right), \tag{8.49}$$

where $\mathcal{O}(\zeta_\theta) \approx 10^{-7}/°K$ and 10^5 Pa $< \mathcal{O}(\zeta_{com}) < 10^{10}$ Pa. The viscosity is assumed to behave according to the well-known relation

$$\frac{\mu_f}{\mu_r} = e^{c(\frac{\theta_r}{\theta_f}-1)}, \tag{8.50}$$

where μ_r is a reference viscosity, θ_r is a reference temperature, and c is a material constant. As before, we introduce the following (per unit mass[2]) decompositions for the key near-field parameters, for example, for the force imparted on particle i by particle j and vice versa:[51]

- $\alpha_{1ij} = \overline{\alpha}_1 m_i m_j$,
- $\alpha_{2ij} = \overline{\alpha}_2 m_i m_j$.

One should expect two primary trends:

- Larger particles are more massive and can impact one another without significant influence from the surrounding fluid. In other words, the particles can "plow" through the fluid and make contact. This makes this situation more thermally volatile, due to the resulting chemical release at contact.
- Smaller particles are more sensitive to the surrounding fluid, and the drag ameliorates the disparity in velocities, thus minimizing the interparticle impact. Thus, these types of flows are less thermally sensitive.

Obviously, in such a model, the number of parameters, even though they are not ad hoc, is large. Thus, corresponding parameter studies would be enormous. This is not the objective of this book. Accordingly, we have taken nominal parameter values that fall roughly in the middle of material data ranges to illustrate the basic approach. The parameters selected for the simulations were as follows:[52]

- a (normalized) domain size of 1 m $\times$1 m $\times$1 m;
- 200 particles randomly distributed in the domain and all started from rest;
- the particle radii randomly distributed in the range $b = 0.05(1 + \pm 0.25)$ m, resulting in approximately 11% of the volume being occupied by the particles;
- an initial velocity of $\boldsymbol{v}_f = (1\text{ m/s}, 0\text{ m/s}, 0\text{ m/s})$ assigned to the fluid and periodic boundary conditions used;
- viscosity parameters $\mu_r = 0.05$ N $-$ s/m^2 and $c = 5$, for the equation of state (Boussinesq-type model), and the same thermal relation assumed for the bulk viscosity, namely, $\frac{\kappa_f}{\kappa_r} = e^{c(\frac{\theta_r}{\theta_f}-1)}$, $\kappa_r = 0.8\mu_r$;[53]
- a uniform initial particle temperature of $\theta = 293.13°$ K;
- a uniform initial fluid interior temperature of $\theta_f = 293.13°$ K serving as the boundary conditions for the domain;
- a particle heat capacity of $C = 1000$ J/(kg °K);

[51] Alternatively, if the near-fields are related to the amount of surface area, this scaling could be done per unit area.

[52] No gravitational effects were considered.

[53] In order to keep the analysis general, we do not enforce the Stokes condition, namely, $\kappa_f = 0$.

- a fluid heat capacity of $C_f = 2500$ J/(kg °K);
- a fluid conductivity of $K_f = 1.0$ Jm^2/(s kg °K);
- a radiative particle emissivity of $\epsilon = 0.05$;
- near-field parameters for the particles of $\overline{\alpha}_1 = 0.1, \overline{\alpha}_2 = 0.01, \beta_1 = 1, \beta_2 = 2$;
- restitution impact coefficients of $e^- = 0.1$ (the lower bound), $e_o = 0.2, \theta^* = 3000°$ K (thermal sensitivity coefficient), $v^* = 10$ m/s;
- a coefficient of static friction of $\mu_s = 0.5$ and a coefficient of dynamic friction of $\mu_d = 0.2$;
- a reaction coefficient of $\xi = 10^9$ J/m^2 and a reaction impact parameter of $I^* = 10^3$ N;
- a heat-drag coefficient of $c_v = 1$;
- a convective heat transfer coefficient of $h_c = 10^3$ J/(sm^2 °K);
- a bulk fluid (compressibility) modulus of $\zeta_{com} = 10^6$ Pa, a reference pressure of $P_o =$ 101300 Pa (1 atm), a reference density of $\rho_o = 1000$ kg/m^3, a reference temperature of $\theta_o = 293.13°$ K, and a thermal expansion coefficient of $\zeta_\theta = 10^{-7°}$ (K)$^{-1}$;
- a particle density of $\rho = 2000$ kg/m^3.

The discretization parameters selected were

- a $10 \times 10 \times 10$ finite difference mesh (with a spacing of 0.1 m) for the numerical derivatives (on the order of the particle size);
- a simulation time of 1 s;
- an initial time step size of 10^{-6} s;
- an upper limit for the time step size of 10^{-2} s;
- a lower limit for the time step size of 10^{-12} s;
- a target number of internal fixed-point iterations of $K_d = 5$;
- a (percentage) iterative (normalized) relative error tolerance within a time step set to $TOL_1 = TOL_2 = TOL_3 = TOL_4 = 10^{-3}$.

8.6 Discussion of the results

For this system, the Reynolds number, based on the mean particle diameter and initial system parameters, was $Re \stackrel{\text{def}}{=} \frac{\rho_o 2 b v_o}{\mu_o} \approx 4010$. The plots in Figures 8.3–8.6 illustrate the system behavior with and without near-fields. There is significant heating due to interparticle collisions when near-fields are present. The presence of near-fields causes particle trajectories due to mutual attraction and repulsion, and particles to make contact frequently.

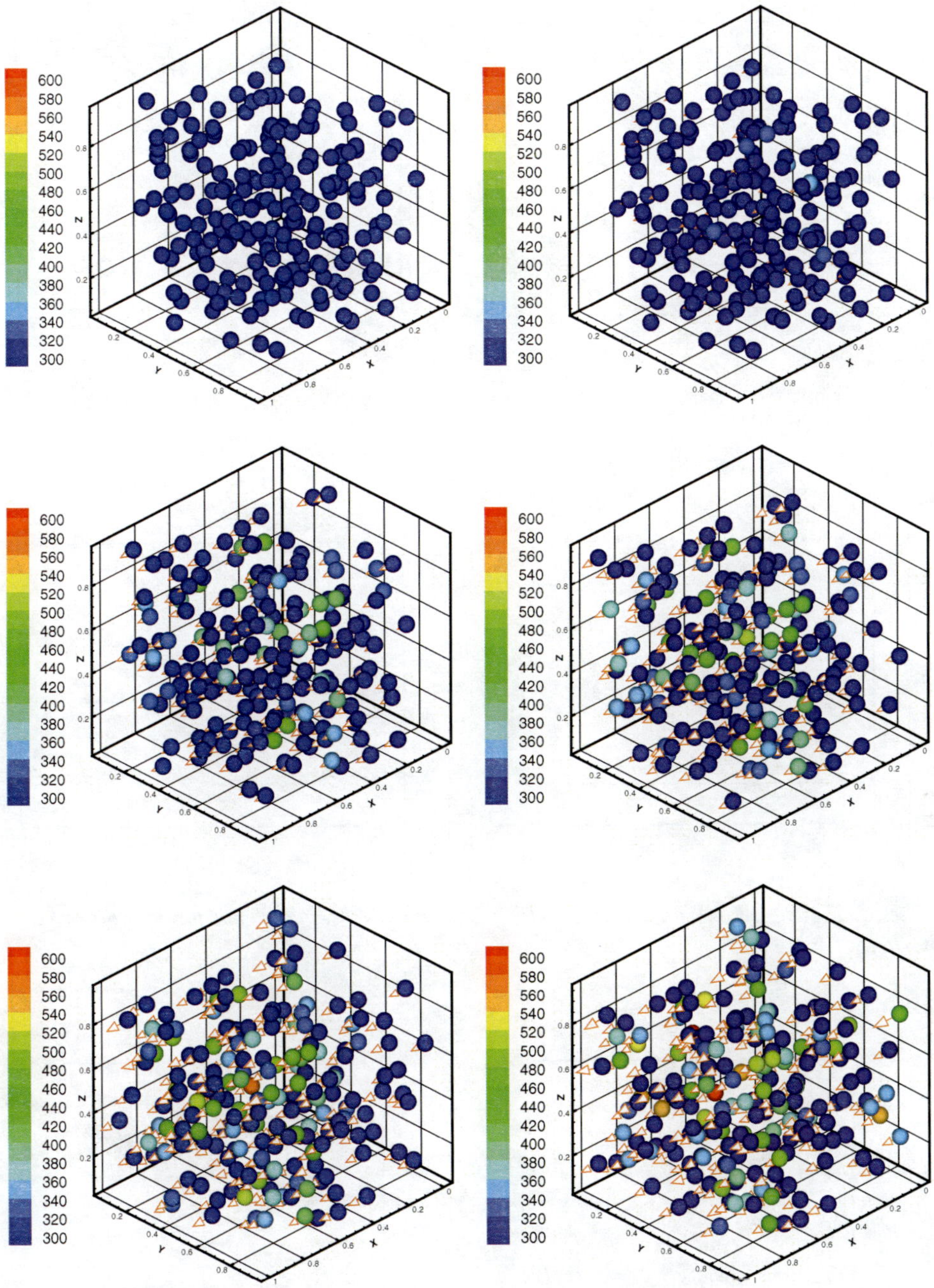

Figure 8.3. *With near-fields: Top to bottom and left to right, the dynamics of the particulate flow. Blue (lowest) indicates a temperature of approximately* 300° K*, while red (highest) indicates a temperature of approximately* 600° K*. The arrows on the particles indicate the velocity vectors (Zohdi* [224]*).*

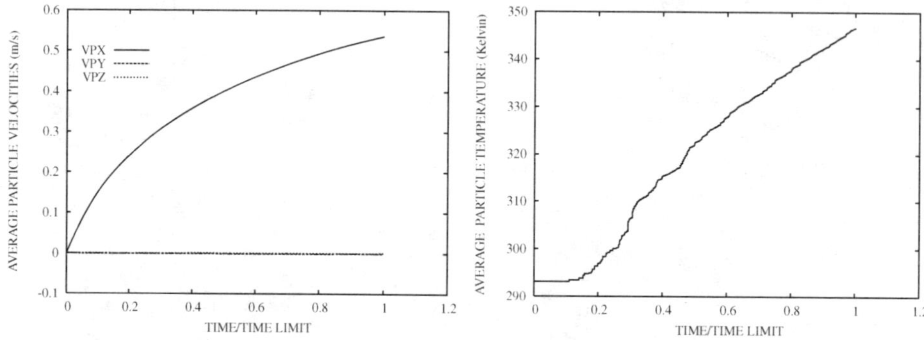

Figure 8.4. *With near-fields: The average velocity and temperature of the particles (Zohdi* [224]*).*

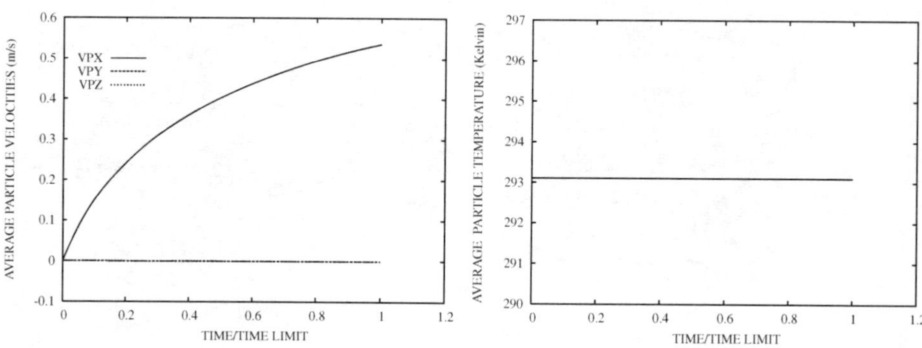

Figure 8.5. *Without near-fields: The average velocity and temperature of the particles (Zohdi* [224]*).*

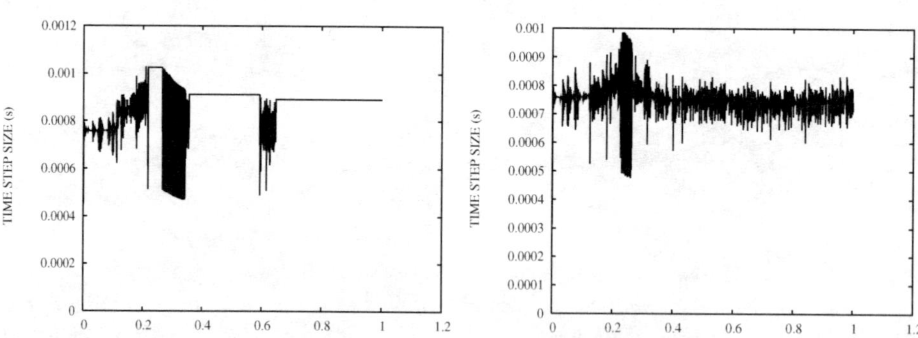

Figure 8.6. *The time step size variation. On the left, with near-fields, and, on the right, without near-fields (Zohdi* [224]*).*

Table 8.1. *Statistics of the particle-laden flow calculations.*

Near-Field	Time Steps	Fixed-Point Iterations	Iter/Time Steps	Time Step Size (s)
Present	1176	8207	6.978	8.506×10^{-4}
Not present	1341	14445	10.772	7.458×10^{-4}

to intersect, In other words, the particles can "plow" through the (compressible) fluid and contact one another. This makes this situation relatively more thermally volatile, due to the resulting chemical release at contact, than cases without near-fields, where the fluid dominates the motion of the particles relatively quickly, not allowing them to make contact. When no near-fields were present, the thermal changes in the particles were negligible, as the plots indicate. A sequence of system configurations are shown in Figure 8.3 for the case where the near-fields are present. Referring to Table 8.1, the total number of time steps needed was 1176 with near-fields and 1342 without near-fields, leading to an average time step size of 8.505×10^{-4} s with near-fields and 7.458×10^{-4} s without near-fields. The number of iterations needed per time step was 6.978 with near-fields and 10.772 without near-fields. We note that while the target iteration limit was set to five iterations per time step, the average value taken for a successful time step exceeds this number, due to the fact that the adaptive algorithm frequently would have to "step back" during the time step refinement process and restart the iterations with a smaller time step. The step sizes varied approximately in the range $4.8 \times 10^{-4} \leq \Delta t \leq 1.1 \times 10^{-3}$ s with near-fields and $4.8 \times 10^{-4} \leq \Delta t \leq 0.9 \times 10^{-3}$ s without near-fields. It is important to note that, in particular for the case with no near-field, time step adaptivity was important throughout the simulation (Figure 8.6). The near-field case's computations converge more quickly. This appears to be due to the fact that when the near-fields are not present, the individual particles have a bit more mobility, and, thus, smaller time steps (slightly more computation) are needed to accurately capture their motion.

8.7 Summary

This work developed a flexible and robust solution strategy to resolve strong multifield coupling between large numbers of particles and a surrounding fluid. As a model problem, a large number of particles undergoing inelastic collisions and simultaneous interparticle (nonlocal) near-field attraction/repulsion were considered. The particles were surrounded by a continuous interstitial fluid that was assumed to obey the fully compressible Navier–Stokes equations. Thermal effects were considered throughout the modeling and simulations. It was assumed that the particles were small enough that the effects of their rotation with respect to their mass centers was unimportant and that any "spin" of the particles was small enough to neglect lift forces that could arise from the interaction with the surrounding fluid. However, the particle-fluid system was strongly coupled due to the drag forces induced by the fluid on the particles and vice versa, as well as the generation of heat due to the drag forces, the thermal softening of the particles, and the thermal dependency of the fluid viscosity. Because the coupling of the various particle and fluid fields can dramatically change over the course of a flow process, the focus of this chapter was on the development of an implicit "staggering" solution scheme, whereby the time steps were adaptively adjusted to control the error associated with the incomplete resolution of the coupled interaction between the various solid particulate and continuum fluid fields. The approach is straightforward and can be easily incorporated into any standard computational fluid mechanics code based on finite difference, finite element, or finite volume discretization. Furthermore, the presented staggering technique, which is designed to resolve the multifield coupling between particles and the surrounding fluid, can be used in a complementary way with other compatible approaches, for example, those developed in the extensive works of Elghobashi and coworkers dealing with particle-laden and bubble-laden fluids (Ferrante and Elghobashi [68], Ahmed and Elghobashi [2], [3], and Druzhinin and Elghobashi [60]). Also, as mentioned earlier, improved descriptions of the fluid-particle interaction can possibly be achieved by using discrete network approximations, which account for hydrodynamic interactions such as those of Berlyand and Panchenko [30] and Berlyand et al. [31].

Chapter 9

Simple optical scattering methods for particulate media

Due to the growing number of applications involving particulate flows, there is a renewed interest in optical detection methods. Ray-tracing is the simplest type of optical model to describe the propagation of light through complex media. The most common physical phenomena associated with rays is in optics, although many other wave phenomena, for example, acoustics, can be described in this manner. *The primary objective here is to introduce the reader to the essential ingredients of classical ray-tracing theory, more appropriately referred to as "geometrical optics," and some modern applications and computational techniques involving particulate media.*[54]

Ray theory is a simple and intuitive approximate theory that can provide sufficiently accurate quantitative information on overall energy propagation for scattering problems in complex media. A further caveat is that ray theory has nearly ideal characteristics for high-performance numerical simulation. For the general state of the art in technical optics, see Gross [86]. For a state of the art review of computational electromagnetics, see Taflove and Hagness [187].

In many instances, the characteristics of flowing, randomly distributed, particulate media are determined by inverse scattering. Essentially, light rays are directed toward the particulate flow, and the characteristics of the particulates, such as their reflectivity and volume fractions, are ascertained by the scattering of the rays. In this chapter, a ray-tracing algorithm is developed and combined with a stochastic genetic algorithm in order to treat coupled inverse optical scattering formulations, where physical parameters, such as particulate volume fractions, refractive indices, and thermal constants, are sought so that the overall response of a sample of randomly distributed particles, suspended in an ambient medium, will match desired coupled scattering, thermal, and infrared responses. Numerical simulations are presented to illustrate the overall procedure and to investigate aggregate ray dynamics corresponding to the flow of electromagnetic energy and the conversion of the absorbed energy into heat and infrared radiation through disordered particulate systems. We shall follow an approach found in Zohdi [218], [219].

[54]Later, we also provide a brief introduction to the field of acoustics and how classical ray theory naturally arises in that field as well.

Remark. It almost goes without saying that the particle positions are assumed fixed relative to the speed of light. In other words, in this chapter the dynamics of the particles plays no role in the analysis.

Remark. We will ignore the phenomenon of diffraction, which originally meant, within the field of optics, a small deviation from rectilinear propagation, but which has come to mean a variety of things to different researchers, for example, the generation of a "shadow" behind a scatterer or the "bending around corners" of incident optical (electromagnetic) waves. It is important to realize that many sophisticated computational methods, which are beyond the scope of this introductory treatment, have geometrical optics, or ray-tracing, as their starting point. Therefore, a clear understanding of ray-tracing is crucial in the study of more advanced methods in optics.

9.1 Introduction

The expressions governing the propagation of electromagnetic waves traveling through space have become known as Maxwell's equations. Virtually all facts about light can be explained in terms of waves.[55] In theory, one could use Maxwell's equations to trace the paths of electromagnetic waves through complex environments. However, when the environment of interest involves hundreds, or thousands, of scatterers, the direct use of Maxwell's equations to describe the flow of energy leads to systems of equations of such complexity that, for all intents and purposes, the problem becomes intractable.

A generally simpler approach is based upon geometrical optics, which makes use of ray-tracing theory and is able to describe various essential aspects of light propagation. This approach is ideal for high-performance computation associated with the scattering of incident light by multiple particles. A variety of applications arise from the reflection and absorption of light in dry particulate flows and related systems comprising randomly dispersed particles suspended in very dilute gases and, in the limit, in a vacuum. For general overviews pertaining to scattering, see Bohren and Huffman [33] and van de Hulst [195].

Remark. An application of particular interest, where scattering calculations can play a supporting role, is the investigation of clustering and aggregation of particles in astrophysical applications where particles collide, cluster, and grow into larger objects. For reviews of such systems, see Chokshi et al. [43], Dominik and Tielens [54], Mitchell and Frenklach [148], Charalampopoulos and Shu [39], [40], and Zohdi [212]–[219].

9.1.1 Ray theory: Scope of use

In this work, we assume that the particle sizes are much greater than the wavelength of the incident light, thus allowing the use of geometrical optics (ray theory). Large particles dictate a way of looking at scattering problems that is quite different from that of scattering due to small particles, where a variety of other techniques are more appropriate (see, for example, Bohren and Huffman [33], Elmore and Heald [63], van de Hulst [195], Hecht [91], Born and Wolf [35], or Gross [86]). In ray theory, an incident beam of light may be thought to consist of separate rays of light, each of which travels along its own path.

[55] Clearly, some effects, such as those pertaining to the momentum transfer of incident light, and the resulting "light pressure," can be explained only in terms of photons (packets of energy).

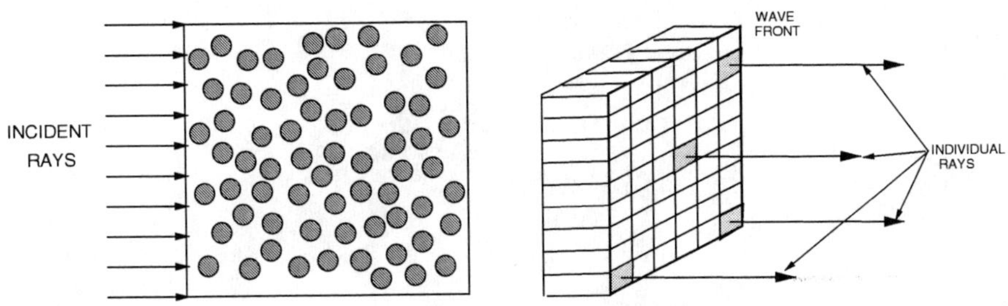

Figure 9.1. *The multiparticle scattering system considered (left), comprised of a beam (right) made up of multiple rays, incident on a collection of randomly distributed scatterers (Zohdi* [218]*).*

Typically, for a particle of radius 10 or more times the size of the wavelength of light, it is possible to distinguish quite clearly between the rays incident on the particle and the rays passing around the particle. Furthermore, experimentally speaking, it is possible to distinguish among rays hitting various parts of the particle's surface. Thus, the rays may be idealized as being localized (Figure 9.1).

One can think of geometrical optics as the limiting case of wave optics where the wavelength (λ) tends toward zero, and as being an approximation to Maxwell's equations, in the same way as Maxwell's equations are an approximation to quantum mechanics models. In other words, classical mechanics is precisely the same limiting approximation to quantum mechanics as geometrical optics is to wave propagation. Essentially, in geometrical optics, the phase of the wave is considered irrelevant. Thus, for ray-tracing to be a valid approach, the wavelengths should be much smaller than those associated with the length scales of the scatterers of the problem at hand (Figure 9.1).

Remark. The wavelengths of visible light fall approximately within 3.8×10^{-7} m $\leq \lambda \leq 7.8 \times 10^{-7}$ m. Note that all electromagnetic radiation travels at the speed of light in a vacuum, $c \approx 3 \times 10^8$ m/s. A more precise value, given by the National Bureau of Standards, is $c \approx 2.997924562 \times 10^8 \pm 1.1$ m/s.

Remark. If the particle sizes are comparable to the wavelength of light, then it is inappropriate to use ray representations. Rayleigh scattering occurs when the scattering particles are smaller than the wavelength of light. Such scattering occurs when light propagates through gases. For example, when sunlight travels through Earth's atmosphere, the light appears to be blue because blue light is more thoroughly scattered than other wavelengths of light. For particle sizes that are on the order of the wavelength of light, the regime is Mie scattering. We do not consider such systems in this work. See Bohren and Huffman [33] and van de Hulst [195] for more details.

9.1.2 Beams composed of multiple rays

In ray-tracing methodology, an incident beam of light, which forms a plane-wave front, which is considered "infinite" in extent (in the lateral directions), relative to the wavelength of light, can be thought of as comprising separate rays of light, each of which pursues its own path. Thus, it almost goes without saying that the width of a beam (w) must satisfy

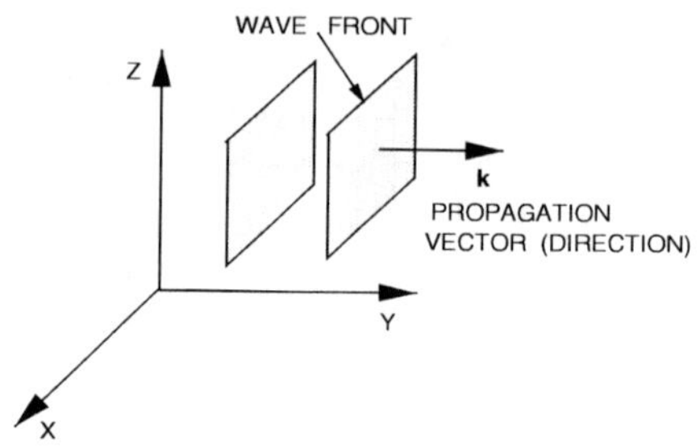

Figure 9.2. *A wave front and propagation vector (Zohdi* [218]*).*

$w \gg \lambda$ for the representation as multiple rays to make sense (Figure 9.1). One can consider the representation of a beam by multiple rays as simply taking a large "sampling" of the diffraction by the beam (wave front) over the portion of the scatterer where the beam is incident. The trajectory of harmonic plane waves, and the corresponding ray representation direction, can actually be derived from Maxwell's equations, which reduce to the classical amplitude and trajectory "Eikonal" equations. For more details, see Born and Wolf [35], Bohren and Huffman [33], Elmore and Heald, [63], and van de Hulst [195].

9.1.3 Objectives

We initially consider coherent beams, representing plane harmonic waves (Figure 9.1), composed of multiple collinear rays, where each ray is a vector in the direction of the flow of electromagnetic energy, which, in isotropic media, corresponds to the normal to the wave front.[56] Thus, for isotropic media, the rays are parallel to the wave's propagation vector, denoted by $\boldsymbol{k}$ (Figure 9.2). Of particular interest is to describe the breakup of initially highly directional coherent beams, which, under normal circumstances, do not spread out into multidirectional rays. A prime example is highly intense light such as that associated with lasers.

In the past, a primary drawback of using a geometrical optics approach has been that it is computationally intensive to track multiple rays, undergoing multiple reflections, energy losses to scatterers, generation of heat, etc. Thus, until relatively recently, the problem of a beam of light, comprising multiple rays, encountering multiple scatterers, has been quite difficult to simulate. However, recent simultaneous advances in numerical methods, coupled with the enormous increase in computational power, have led to the possibility that such problems are accessible to rapid desktop computing. Accordingly, in this chapter a ray-tracing algorithm is developed and combined with a stochastic genetic algorithm in order to treat coupled inverse optical scattering formulations, where physical parameters, such as particulate volume fractions, refractive indices, and thermal constants, are sought so that the overall response of a sample of randomly distributed suspensions will match desired scattering, thermal, and infrared responses. Numerical simulations are presented to illustrate the overall procedure and to investigate aggregate ray dynamics corresponding to the flow of electromagnetic energy and the conversion of the absorbed energy into heat and infrared radiation through disordered particulate systems.

[56] Beams consisting of parallel rays are sometimes referred to as "collimated" light beams.

9.2 Plane harmonic electromagnetic waves

9.2.1 Plane waves

We recall the basic form of the wave equation

$$\frac{\partial^2 A}{\partial x^2} + \frac{\partial^2 A}{\partial y^2} + \frac{\partial^2 A}{\partial z^2} = \frac{1}{v^2}\frac{\partial^2 A}{\partial t^2}, \tag{9.1}$$

where A is a variable and v is the wave speed. We consider time-harmonic plane wave solutions, i.e., those solutions of the form

$$A(\boldsymbol{r}, t) = A_o \cos(\boldsymbol{k} \cdot \boldsymbol{r} - \omega t), \tag{9.2}$$

where $\boldsymbol{r}$ is an initial position vector to the wave front and $\boldsymbol{k}$ is in the direction of propagation. For plane waves $\boldsymbol{k} \cdot \boldsymbol{r} =$ const. We denote the phase as

$$\phi = \boldsymbol{k} \cdot \boldsymbol{r} - \omega t \tag{9.3}$$

and the angular frequency as $\omega = \frac{2\pi}{\tau}$, where τ is the period. The wave front, over which the phase is constant, is a plane for "plane waves" and is orthogonal to the direction of propagation.

9.2.2 Electromagnetic waves

As we have indicated, the propagation of light can be described via an electromagnetic formalism, Maxwell's equations (in simplified form), in free space:

$$\nabla \times \boldsymbol{E} = -\mu_o \frac{\partial \boldsymbol{H}}{\partial t}, \quad \nabla \times \boldsymbol{H} = \epsilon_o \frac{\partial \boldsymbol{E}}{\partial t}, \quad \nabla \cdot \boldsymbol{H} = 0, \quad \text{and} \quad \nabla \cdot \boldsymbol{E} = 0, \tag{9.4}$$

where $\boldsymbol{E}$ is the electric field intensity, $\boldsymbol{H}$ is the magnetic flux intensity, ϵ_o is the permittivity, and μ_o is the permeability. Using standard vector identities, one can show that

$$\nabla \times (\nabla \times \boldsymbol{E}) = -\mu_o \epsilon_o \frac{\partial^2 \boldsymbol{E}}{\partial t^2} \quad \text{and} \quad \nabla \times (\nabla \times \boldsymbol{H}) = -\mu_o \epsilon_o \frac{\partial^2 \boldsymbol{H}}{\partial t^2}, \tag{9.5}$$

that

$$\nabla^2 \boldsymbol{E} = \frac{1}{c^2}\frac{\partial^2 \boldsymbol{E}}{\partial t^2} \quad \text{and} \quad \nabla^2 \boldsymbol{H} = \frac{1}{c^2}\frac{\partial^2 \boldsymbol{H}}{\partial t^2}, \tag{9.6}$$

and that, employing a Cartesian coordinate system,

$$\frac{\partial^2 E_x}{\partial x^2} + \frac{\partial^2 E_x}{\partial y^2} + \frac{\partial^2 E_x}{\partial z^2} = \frac{1}{c^2}\frac{\partial^2 E_x}{\partial t^2}, \tag{9.7}$$

where $c = \frac{1}{\sqrt{\epsilon_o \mu_o}}$, with identical relations holding for E_y, E_z, H_x, H_y, and H_z. In the case of plane harmonic waves, for example, of the form

$$\boldsymbol{E} = \boldsymbol{E}_o \cos(\boldsymbol{k} \cdot \boldsymbol{r} - \omega t) \quad \text{and} \quad \boldsymbol{H} = \boldsymbol{H}_o \cos(\boldsymbol{k} \cdot \boldsymbol{r} - \omega t), \tag{9.8}$$

we have

$$\boldsymbol{k} \times \boldsymbol{E} = \mu_o \omega \boldsymbol{H} \quad \text{and} \quad \boldsymbol{k} \times \boldsymbol{H} = -\epsilon_o \omega \boldsymbol{E} \tag{9.9}$$

and

$$\boldsymbol{k} \cdot \boldsymbol{E} = 0 \quad \text{and} \quad \boldsymbol{k} \cdot \boldsymbol{H} = 0. \tag{9.10}$$

Vectors, $\boldsymbol{k}$, $\boldsymbol{E}$, and $\boldsymbol{H}$ form a mutually orthogonal triad. The direction of ray propagation is given by $\frac{\boldsymbol{E}\times\boldsymbol{H}}{\|\boldsymbol{E}\times\boldsymbol{H}\|}$. Since the free-space propagation velocity is given by $c = \frac{1}{\sqrt{\epsilon_o \mu_o}}$ for an electromagnetic wave in a vacuum and $v = \frac{1}{\sqrt{\epsilon\mu}}$ for electromagnetic waves in another medium, we can define the index of refraction as

$$n \stackrel{\text{def}}{=} \frac{c}{v} = \sqrt{\frac{\epsilon\mu}{\epsilon_o \mu_o}}. \tag{9.11}$$

9.2.3 Optical energy propagation

Light waves traveling through space carry electromagnetic energy that flows in the direction of wave propagation. The energy per unit area per unit time flowing perpendicularly into a surface in free space is given by the Poynting vector $\boldsymbol{S}$, where

$$\boldsymbol{S} = \boldsymbol{E} \times \boldsymbol{H}. \tag{9.12}$$

Since at optical frequencies $\boldsymbol{E}$, $\boldsymbol{H}$, and $\boldsymbol{S}$ oscillate rapidly, it is impractical to measure instantaneous values of $\boldsymbol{S}$ directly. Now consider the harmonic representations in Equation (9.8), which lead to

$$\boldsymbol{S} = \boldsymbol{E}_o \times \boldsymbol{H}_o \cos^2(\boldsymbol{k} \cdot \boldsymbol{r} - \omega t) \tag{9.13}$$

and, consequently, the average value over a longer (but still quite short) time interval than that of the time scale of rapid random oscillation,

$$\langle \boldsymbol{S} \rangle_{\mathcal{T}} = \boldsymbol{E}_o \times \boldsymbol{H}_o \langle \cos^2(\boldsymbol{k} \cdot \boldsymbol{r} - \omega t) \rangle_{\mathcal{T}} = \frac{1}{2} \boldsymbol{E}_o \times \boldsymbol{H}_o, \tag{9.14}$$

where $\langle (\cdot) \rangle_{\mathcal{T}} \stackrel{\text{def}}{=} \frac{1}{\mathcal{T}} \int_0^{\mathcal{T}} (\cdot)\, dt$. We define the irradiance as

$$I \stackrel{\text{def}}{=} \langle \|\boldsymbol{S}\| \rangle_{\mathcal{T}} = \frac{1}{2} \|\boldsymbol{E}_o \times \boldsymbol{H}_o\| = \frac{1}{2} \sqrt{\frac{\epsilon_o}{\mu_o}} \|\boldsymbol{E}_o\|^2. \tag{9.15}$$

Clearly, the rate of flow of energy is proportional to the square of the amplitude of the electric field and, in isotropic media, which we consider for the duration of the work, the flow of energy moves in the direction of $\boldsymbol{S}$ and in the same direction as $\boldsymbol{k}$. Since I is the energy per unit area per unit time, if we multiply by the "cross-sectional" area of the ray, a_r, we obtain the energy associated with the ray, denoted as $I a_r = I a_b / N_r$, where a_b is the cross-sectional area of a beam (comprising all of the rays) and N_r is the number of rays in the beam (Figure 9.3). A concise introduction can be found in Fowles [70].

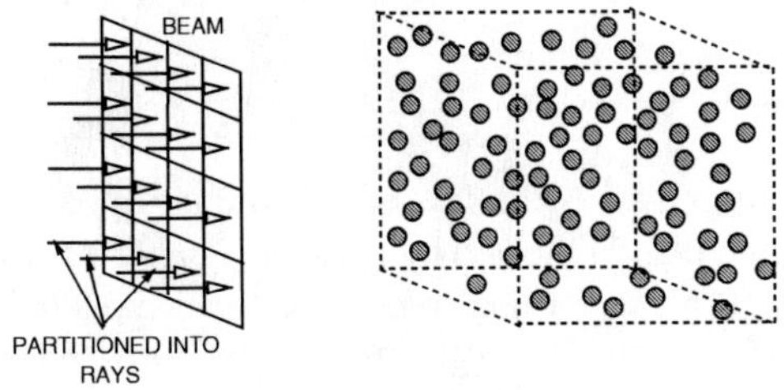

Figure 9.3. *The scattering system considered, comprising a beam made up of multiple rays, incident on a collection of randomly distributed scatterers.*

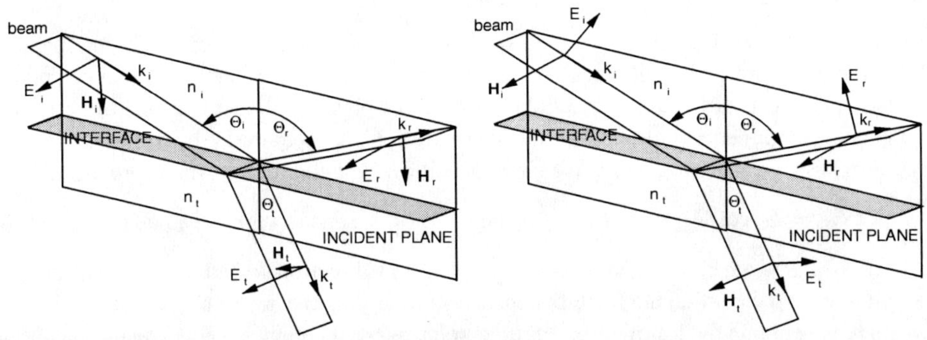

Figure 9.4. *The nomenclature for Fresnel's equations, for the case where the electric field vectors are (left) perpendicular to the plane of incidence and (right) parallel to the plane of incidence (Zohdi* [218]*).*

9.2.4 Reflection and absorption of energy

Now we consider a plane harmonic wave incident upon a plane boundary (material interface) separating two optically different materials, which produces a reflected wave and a transmitted (refracted) wave (Figure 9.4). The space-time dependence of the three waves is given by (1) $e^{j(\boldsymbol{k}_i\cdot\boldsymbol{r}-\omega t)}$ for the incident wave (with propagation vector $\boldsymbol{k}_i$), (2) $e^{j(\boldsymbol{k}_r\cdot\boldsymbol{r}-\omega t)}$ for the reflected wave (with propagation vector $\boldsymbol{k}_r$), and (3) $e^{j(\boldsymbol{k}_t\cdot\boldsymbol{r}-\omega t)}$ for the transmitted wave (with propagation vector $\boldsymbol{k}_t$). In order for a time-invariant relation to hold for all points on the boundary, and for all values of t, we must have that the arguments of the exponential function are equal on the boundary. Therefore, since the ωt terms are the same, we have, at the boundary, $\boldsymbol{k}_i \cdot \boldsymbol{r} = \boldsymbol{k}_r \cdot \boldsymbol{r} = \boldsymbol{k}_t \cdot \boldsymbol{r}$, which implies that the waves are coplanar and that their projection onto the plane boundary is equal. We call the plane that contains all three waves the incident plane. Consequently, we have a relation between the propagation constants' magnitudes, $k_i \sin\theta_i = k_r \sin\theta_r = k_t \sin\theta_t$, which implies, because the reflected and incident medium are the same, $\theta_i = \theta_r$. By taking the ratio of the magnitudes of the propagation constants of the transmitted wave and the incident wave, we have

$$\frac{k_t}{k_i} = \frac{\omega/v_t}{\omega/v_i} = \frac{c/v_t}{c/v_i} = \frac{n_t}{n_i} \stackrel{\text{def}}{=} \hat{n}. \tag{9.16}$$

Therefore, we have

$$\frac{\sin\theta_i}{\sin\theta_t} = \hat{n}, \tag{9.17}$$

which is sometimes referred to as the law of refraction. To compute the amount of energy transmitted (absorbed) and reflected by electromagnetic waves, let $\boldsymbol{E}_i$ now denote the (vectorial) amplitude of a plane harmonic wave that is incident on a plane boundary separating two materials. Also, let $\boldsymbol{E}_r$ and $\boldsymbol{E}_t$ be the amplitudes of the reflected and transmitted waves, respectively. Equations (9.9) and (9.10) collapse to, for the incident, reflected, and transmitted magnetic waves,

$$\boldsymbol{H}_i = \frac{1}{\mu_i \omega}\boldsymbol{k}_i \times \boldsymbol{E}_i, \qquad \boldsymbol{H}_r = \frac{1}{\mu_r \omega}\boldsymbol{k}_r \times \boldsymbol{E}_r, \qquad \boldsymbol{H}_t = \frac{1}{\mu_t \omega}\boldsymbol{k}_t \times \boldsymbol{E}_t. \tag{9.18}$$

Let us now consider an oblique angle of incidence. Consider two cases for the electric field vector: (1) electric field vectors that are parallel ($||$) to the plane of incidence and (2) electric field vectors that are perpendicular ($\perp$) to the plane of incidence. In either case, the tangential components of the electric and magnetic fields are required to be continuous across the interface. Consider case (1). We have the following general vectorial representations:

$$\boldsymbol{E}_{||} = E_{||}\cos(\boldsymbol{k}\cdot\boldsymbol{r} - \omega t)\,\boldsymbol{e}_1 \qquad \text{and} \qquad \boldsymbol{H}_{||} = H_{||}\cos(\boldsymbol{k}\cdot\boldsymbol{r} - \omega t)\,\boldsymbol{e}_2, \tag{9.19}$$

where $\boldsymbol{e}_1$ and $\boldsymbol{e}_2$ are orthogonal to the propagation direction $\boldsymbol{k}$ and $E_{||}$ and $H_{||}$ are the amplitudes of the parallel field components. By employing the law of refraction ($n_i \sin\theta_i = n_t \sin\theta_t$), we obtain the following conditions relating the incident, reflected, and transmitted components of the electric field quantities:

$$E_{||i}\cos\theta_i - E_{||r}\cos\theta_r = E_{||t}\cos\theta_t \qquad \text{and} \qquad H_{\perp i} + H_{\perp r} = H_{\perp t}. \tag{9.20}$$

Since, for plane harmonic waves, the magnetic and electric field amplitudes are related by $H = \frac{E}{v\mu}$, we then have

$$E_{||i} + E_{||r} = \frac{\mu_i}{\mu_t}\frac{v_i}{v_t}E_{||t} = \frac{\mu_i}{\mu_t}\frac{n_t}{n_i}E_{||t} \stackrel{\text{def}}{=} \frac{\hat{n}}{\hat{\mu}}E_{||t}, \tag{9.21}$$

where $\hat{\mu} \stackrel{\text{def}}{=} \frac{\mu_t}{\mu_i}$, $\hat{n} \stackrel{\text{def}}{=} \frac{n_t}{n_i}$, and v_i, v_r, and v_t are the values of the velocity in the incident, reflected, and transmitted directions.[57] By again employing the law of refraction, we obtain the Fresnel reflection and transmission coefficients, generalized for the case of unequal magnetic permeabilities:

$$r_{||} = \frac{E_{||r}}{E_{||i}} = \frac{\frac{\hat{n}}{\hat{\mu}}\cos\theta_i - \cos\theta_t}{\frac{\hat{n}}{\hat{\mu}}\cos\theta_i + \cos\theta_t} \qquad \text{and} \qquad t_{||} = \frac{E_{||t}}{E_{||i}} = \frac{2\cos\theta_i}{\cos\theta_t + \frac{\hat{n}}{\hat{\mu}}\cos\theta_i}. \tag{9.22}$$

Following the same procedure for case (2), where the components of $\boldsymbol{E}$ are perpendicular to the plane of incidence, we have

$$r_{\perp} = \frac{E_{\perp r}}{E_{\perp i}} = \frac{\cos\theta_i - \frac{\hat{n}}{\hat{\mu}}\cos\theta_t}{\cos\theta_i + \frac{\hat{n}}{\hat{\mu}}\cos\theta_t} \qquad \text{and} \qquad t_{\perp} = \frac{E_{\perp t}}{E_{\perp i}} = \frac{2\cos\theta_i}{\cos\theta_i + \frac{\hat{n}}{\hat{\mu}}\cos\theta_t}. \tag{9.23}$$

[57] Throughout the analysis we assume that $\hat{n} \geq 1$.

Our primary interest is in the reflections. We define the reflectances as

$$R_{\|} \stackrel{\text{def}}{=} r_{\|}^2 \qquad \text{and} \qquad R_{\perp} \stackrel{\text{def}}{=} r_{\perp}^2. \tag{9.24}$$

Particularly convenient forms for the reflections are

$$r_{\|} = \frac{\frac{\hat{n}^2}{\hat{\mu}}\cos\theta_i - (\hat{n}^2 - \sin^2\theta_i)^{\frac{1}{2}}}{\frac{\hat{n}^2}{\hat{\mu}}\cos\theta_i + (\hat{n}^2 - \sin^2\theta_i)^{\frac{1}{2}}} \qquad \text{and} \qquad r_{\perp} = \frac{\cos\theta_i - \frac{1}{\hat{\mu}}(\hat{n}^2 - \sin^2\theta_i)^{\frac{1}{2}}}{\cos\theta_i + \frac{1}{\hat{\mu}}(\hat{n}^2 - \sin^2\theta_i)^{\frac{1}{2}}}. \tag{9.25}$$

Thus, the total energy reflected can be characterized by

$$R \stackrel{\text{def}}{=} \left(\frac{E_r}{E_i}\right)^2 = \frac{E_{\perp r}^2 + E_{\|r}^2}{E_i^2} = \frac{I_{\|r} + I_{\perp r}}{I_i}. \tag{9.26}$$

If the resultant plane of oscillation of the (polarized) wave makes an angle of γ_i with the plane of incidence, then

$$E_{\|i} = E_i \cos\gamma_i \qquad \text{and} \qquad E_{\perp i} = E_i \sin\gamma_i, \tag{9.27}$$

and it follows from the previous definition of I that

$$I_{\|i} = I_i \cos^2\gamma_i \qquad \text{and} \qquad I_{\perp i} = I_i \sin^2\gamma_i. \tag{9.28}$$

Substituting these expressions back into the expressions for the reflectances yields

$$R = \frac{I_{\|r}}{I_{\|i}}\cos^2\gamma_i + \frac{I_{\perp r}}{I_{\|i}}\sin^2\gamma_i = R_{\|}\cos^2\gamma_i + R_{\perp}\sin^2\gamma_i. \tag{9.29}$$

For natural or unpolarized light, the angle γ_i varies rapidly in a random manner, as does the field amplitude. Thus, since

$$\langle\cos^2\gamma_i(t)\rangle_T = \frac{1}{2} \qquad \text{and} \qquad \langle\sin^2\gamma_i(t)\rangle_T = \frac{1}{2}, \tag{9.30}$$

and therefore for natural light

$$I_{\|i} = \frac{I_i}{2} \qquad \text{and} \qquad I_{\perp i} = \frac{I_i}{2}, \tag{9.31}$$

we have

$$r_{\|}^2 = \left(\frac{E_{\|r}^2}{E_{\|i}^2}\right)^2 = \frac{I_{\|r}}{I_{\|i}} \qquad \text{and} \qquad r_{\perp}^2 = \left(\frac{E_{\perp r}^2}{E_{\perp i}^2}\right)^2 = \frac{I_{\perp r}}{I_{\perp i}}. \tag{9.32}$$

Thus, the total reflectance becomes

$$R = \frac{1}{2}(R_{\|} + R_{\perp}) = \frac{1}{2}(r_{\|}^2 + r_{\perp}^2), \tag{9.33}$$

where $0 \leq R \leq 1$.

Remark. For the cases where $\sin\theta_t = \frac{\sin\theta_i}{\hat{n}} > 1$, one may rewrite the reflection relations as

$$r_{||} = \frac{\frac{\hat{n}^2}{\hat{\mu}}\cos\theta_i - j(\sin^2\theta_i - \hat{n}^2)^{\frac{1}{2}}}{\frac{\hat{n}^2}{\hat{\mu}}\cos\theta_i + j(\sin^2\theta_i - \hat{n}^2)^{\frac{1}{2}}} \quad \text{and} \quad r_{\perp} = \frac{\cos\theta_i - \frac{1}{\hat{\mu}}j(\sin^2\theta_i - \hat{n}^2)^{\frac{1}{2}}}{\cos\theta_i + \frac{1}{\hat{\mu}}j(\sin^2\theta_i - \hat{n}^2)^{\frac{1}{2}}}, \tag{9.34}$$

where $j = \sqrt{-1}$ and, in this complex case,[58]

$$R_{||} \stackrel{\text{def}}{=} r_{||}\bar{r}_{||} = 1 \quad \text{and} \quad R_{\perp} \stackrel{\text{def}}{=} r_{\perp}\bar{r}_{\perp} = 1, \tag{9.35}$$

where $\bar{r}_{||}$ and $\bar{r}_{\perp}$ are complex conjugates. Thus, for angles above the critical angle θ_i^*, all of the energy is reflected.

Remark. Notice that as $\hat{n} \to 1$ we have complete absorption, while as $\hat{n} \to \infty$ we have complete reflection. The total amount of absorbed power by the particles is $(1 - R)I_i$. As mentioned previously, the medium surrounding the particles is assumed to behave as a vacuum, i.e., there are no energetic losses as the electromagnetic rays pass through it. However, we assume that all electromagnetic energy that is absorbed from a ray by a particle is converted into heat and that no electromagnetic rays are refracted or dispersed. Heat generation and accompanying thermal radiation emission (with wavelengths in the range of approximately 10^{-7} m $\leq \lambda \leq 10^{-4}$ m) are addressed next.

Remark. The amount of incident electromagnetic energy (I_i) that is reflected (I_r) is given by the total reflectance (Figure 9.5)

$$R \stackrel{\text{def}}{=} \frac{I_r}{I_i}, \tag{9.36}$$

where $0 \leq R \leq 1$ and where, explicitly for unpolarized (natural) light,

$$R = \frac{1}{2}\left(\left(\frac{\frac{\hat{n}^2}{\hat{\mu}}\cos\theta_i - (\hat{n}^2 - \sin^2\theta_i)^{\frac{1}{2}}}{\frac{\hat{n}^2}{\hat{\mu}}\cos\theta_i + (\hat{n}^2 - \sin^2\theta_i)^{\frac{1}{2}}}\right)^2 + \left(\frac{\cos\theta_i - \frac{1}{\hat{\mu}}(\hat{n}^2 - \sin^2\theta_i)^{\frac{1}{2}}}{\cos\theta_i + \frac{1}{\hat{\mu}}(\hat{n}^2 - \sin^2\theta_i)^{\frac{1}{2}}}\right)^2\right). \tag{9.37}$$

For most materials, the magnetic permeability is, within experimental measurements, virtually the same.[59] For the remainder of the work, we shall take $\hat{\mu} = 1$, i.e., $\mu_o = \mu_i \approx \mu_t$. However, further comments on the sensitivity of the reflectance to $\hat{\mu}$ are given later, in the concluding comments and in Appendix B.

Remark. In the upcoming analysis, the ambient medium is assumed to behave as a vacuum. Thus, there are no energetic losses as the electromagnetic rays pass through it. However, we assume that all electromagnetic energy that is absorbed by a particle becomes trapped, and not re-emitted. Such energy is assumed to be converted into heat. The thermal conversion process, and subsequent infrared radiation emission, is not considered in the present work. Modeling of the thermal coupling in such processes can be found in Zohdi [218] and will be described later in detail. Thus, we ignore the transmission of

---

[58] The limiting case $\frac{\sin\theta_i^*}{\hat{n}} = 1$ is the critical angle (θ_i^*) case.

[59] A few notable exceptions are concentrated magnetite, pyrrhotite, and titanomagnetite (Telford et al. [192] and Nye [153]).

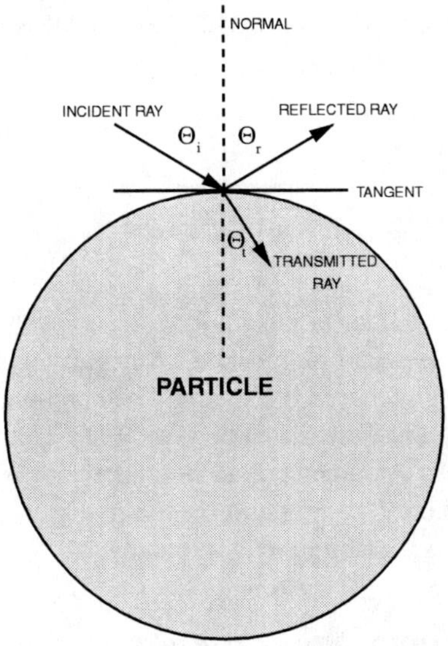

Figure 9.5. *The nomenclature for Fresnel's equations for a incident ray that encounters a scattering particle (Zohdi* [219]*).*

light through the scattering particles, as well as dispersion, i.e., the decomposition of light into its component wavelengths (or colors). This phenomenon occurs because the index of refraction of a transparent medium is greater for light of shorter wavelengths. Thus, whenever light is refracted in passing from one medium to the next, the violet and blue light of shorter wavelengths is bent more than the orange and red light of longer wavelengths. Dispersive effects introduce a new level of complexity, primarily because of the refraction of different wavelengths of light, leading to a dramatic growth in the number of rays of varying intensities and color (wavelength).

9.3 Multiple scatterers

The primary quantity of interest in this work is the percentage of "lost" irradiance by a beam encountering a collection of randomly distributed particles in a selected direction over the time interval of $(0, T)$. This is characterized by the inner product of the Poynting vector and a selected direction ($\boldsymbol{d}$):

$$\mathcal{Z}(0,T) \stackrel{\text{def}}{=} \frac{\sum_{i=1}^{N_r}(\boldsymbol{S}_i(t=0) - \boldsymbol{S}_i(t=T)) \cdot \boldsymbol{d}}{\sum_{i=1}^{N_r} \boldsymbol{S}_i(t=0) \cdot \boldsymbol{d}}, \tag{9.38}$$

where $\mathcal{Z}$ can be considered the amount of energy "blocked" (in a vectorially averaged sense) from propagating in the $\boldsymbol{d}$ direction. Now consider a cost function comparing the loss to the specified blocked amount:

$$\Pi \stackrel{\text{def}}{=} \frac{\mathcal{Z}(0,T) - \mathcal{Z}^*}{\mathcal{Z}^*}, \tag{9.39}$$

($\star$) COMPUTE RAY ORIENTATIONS AFTER REFLECTION (FRESNEL RELATIONS);
COMPUTE ABSORPTION BY PARTICLES;
INCREMENT ALL RAY POSITIONS: $\boldsymbol{r}_i(t+\Delta t) = \boldsymbol{r}_i(t) + \Delta t \boldsymbol{v}_i(t), \quad i = 1, \ldots, RAYS$;
GO TO ($\star$) *AND REPEAT WITH* $t = t + \Delta t$.

ALGORITHM 9.1

where $\mathcal{Z}^*$ is a target blocked value. For example, if $\mathcal{Z}^* = 1$, then we want all of the energy, *in a vectorially averaged sense*, to be blocked. A negative value of Π means that, in an overall sense, rays are being scattered backward. The computational algorithm, Algorithm 9.1, is given above, starting at $t = 0$ and ending at $t = T$. The time step size Δt is dictated by the size of the particles. A somewhat ad hoc approach is to scale the time step size according to $\Delta t \propto \frac{\xi b}{||\boldsymbol{v}||}$, where b is the radius of the particles, $||\boldsymbol{v}||$ is the magnitude of the velocity of the rays, and ξ is a scaling factor, typically $0.05 \leq \xi \leq 0.1$.

9.3.1 Parametrization of the scatterers

We considered a group of N_p randomly positioned particles, of equal size, in a cubical domain of dimensions $D \times D \times D$, where $D = 10^{-3}$ m. The particle size and volume fraction were determined by a particle/sample size ratio, which was defined via a subvolume size $V \stackrel{\text{def}}{=} \frac{D \times D \times D}{N_p}$, where N_p was the number of particles in the entire cube. The ratio between the radius (b) and the subvolume was denoted by $\mathcal{L} \stackrel{\text{def}}{=} \frac{b}{V^{\frac{1}{3}}}$. The volume fraction occupied by the particles can consequently be written as $v_p \stackrel{\text{def}}{=} \frac{4\pi \mathcal{L}^3}{3}$. Thus, the total volume occupied by the particles,[60] denoted by ζ, can be written as $\zeta = v_f N_p V$. We used $N_p = 1000$ particles and $N_r = 400$ rays, arranged in a square 20×20 pattern (Figure 9.6). This system provided stable results, i.e., increasing the number of rays and/or the number of particles beyond these levels resulted in negligibly different overall system responses. The irradiance beam parameter was set to $I = 10^{18}$ J/(m$^2\cdot$ s), where the irradiance for each ray was calculated as $I a_b / N_r$, where $N_r = 20 \times 20 = 400$ is the number of rays in the beam and $a_b = 10^{-3}$ m $\times 10^{-3}$ m $= 10^{-6}$ m^2 is the cross-sectional area of the beam.[61] The simulations were run until the rays completely exited the domain, which corresponded to a time scale on the order of $\frac{3 \times 10^{-3}\text{ m}}{c}$, where c is the speed of light. The initial velocity vector for all of the initially collinear rays making up the beam was $\boldsymbol{v} = (c, 0, 0)$. The particle length scale $\mathcal{L}$ was varied between 0.25 and 0.375, while the relative refractive index ratio $\hat{n}$ was varied between 2 and 100.

Remark. Typically, for a random realization of scatterers, comprising a finite number of particles, there will be slight variations in the response (Π) for different random configurations. In order to stabilize Π's value with respect to the randomness for a given parameter selection, comprising particle length scales, relative refractive indices, etc., denoted by

[60] For example, if one were to arrange the particles in a regular periodic manner, then at the length scale ratio of $\mathcal{L} = 0.25$ the distance between the centers of the particle becomes four particle radii. In theoretical works, it is often stated that the critical separation distance between particles is approximately three radii to be sufficient to treat the particles as independent scatterers and simply to sum the effects of the individual scatterers to compute the overall response of the aggregate.

[61] Because of the normalized structure of the blocking function, Π, it is insensitive to the magnitude of I.

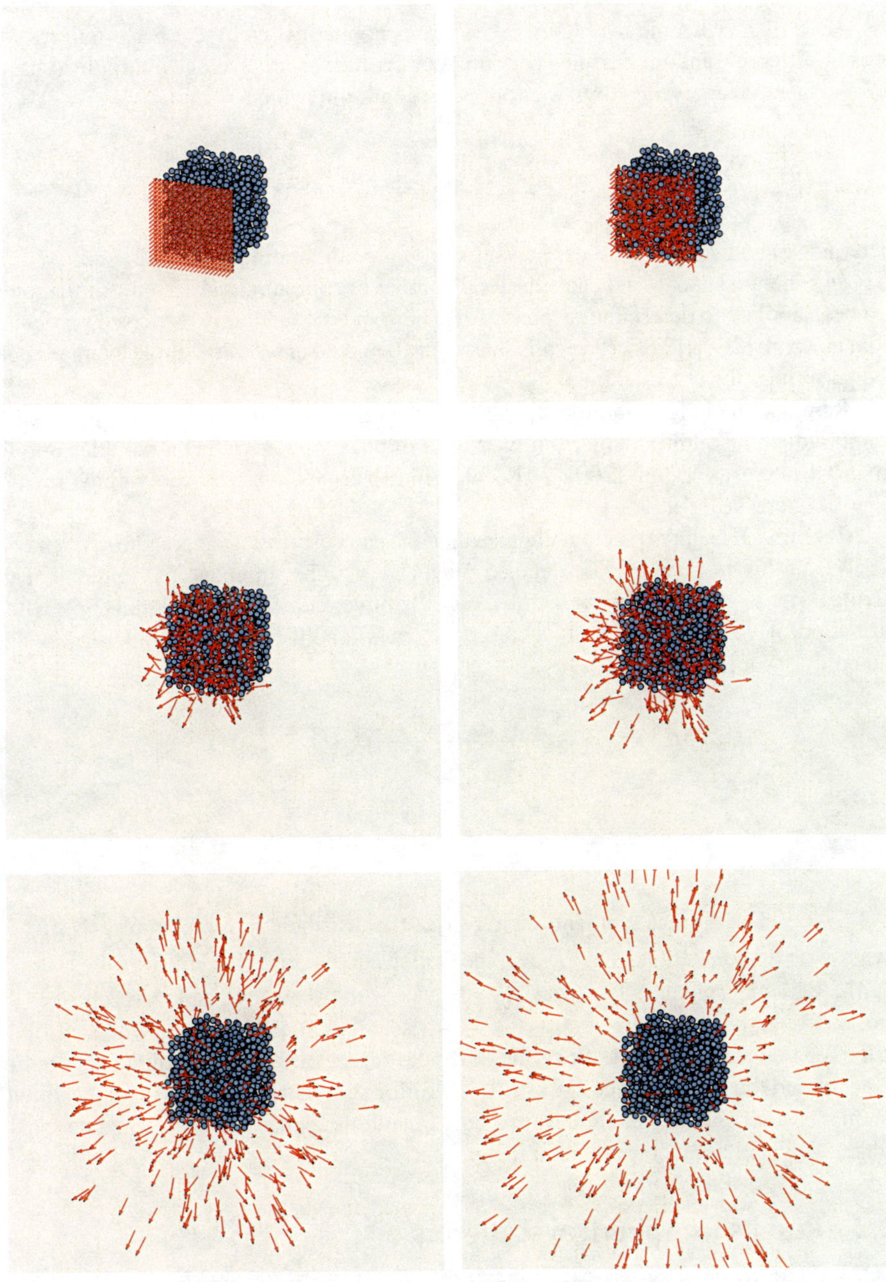

Figure 9.6. *Top to bottom and left to right, the progressive movement of rays making up a beam ($\mathcal{L} = 0.325$ and $\hat{n} = 10$). The lengths of the vectors indicate the irradiance (Zohdi* [219]*).*

$\mathbf{\Lambda} \stackrel{\text{def}}{=} (\mathcal{L}, \hat{n})$, an ensemble averaging procedure is applied whereby the performances of a series of different random starting scattering configurations are averaged until the (ensemble) average converges, i.e., until the following condition is met:

$$\left|\frac{1}{M+1}\sum_{i=1}^{M+1}\Pi^{(i)}(\mathbf{\Lambda}^I) - \frac{1}{M}\sum_{i=1}^{M}\Pi^{(i)}(\mathbf{\Lambda}^I)\right| \leq TOL\left|\frac{1}{M+1}\sum_{i=1}^{M+1}\Pi^{(i)}(\mathbf{\Lambda}^I)\right|,$$

where index i indicates a different starting random configuration ($i = 1, 2, \ldots, M$) that has been generated and M indicates the total number of configurations tested. Similar ideas have been applied to determine responses of other types of randomly dispersed particulate media in Zohdi [208]–[213]. Typically, between 10 and 20 ensemble sample averages need to be performed for Π to stabilize.

Remark. As before, in order to generate the random particle positions, the classical random sequential addition algorithm was used to place nonoverlapping particles into the domain of interest (Widom [200]). This algorithm was adequate for the volume fraction ranges of interest (under 30%).

Remark. It is important to recognize that one can describe the aggregate ray behavior described in this work in a more detailed manner via higher moment distributions of the individual ray fronts and their velocities. For example, consider any quantity, Q, with a distribution of values ($Q_i, i = 1, 2, \ldots, N_r$ = rays) about an arbitrary reference value, denoted $Q^\star$, as follows:

$$\boldsymbol{M}_p^{Q_i - Q^\star} \stackrel{\text{def}}{=} \frac{\sum_{i=1}^{N_r}(Q_i - Q^\star)^p}{N_r} \stackrel{\text{def}}{=} \overline{(Q_i - Q^\star)^p}, \tag{9.40}$$

where

$$\frac{\sum_{i=1}^{N_r}(\cdot)}{N_r} \stackrel{\text{def}}{=} \overline{(\cdot)} \tag{9.41}$$

and $A \stackrel{\text{def}}{=} \overline{Q_i}$. The various moments characterize the distribution, for example, (I) $\boldsymbol{M}_1^{Q_i-A}$ measures the first deviation from the average, which equals zero, (II) $\boldsymbol{M}_1^{Q_i-0}$ is the average, (III) $\boldsymbol{M}_2^{Q_i-A}$ is the standard deviation, (IV) $\boldsymbol{M}_3^{Q_i-A}$ is the skewness, and (V) $\boldsymbol{M}_4^{Q_i-A}$ is the kurtosis. The higher moments, such as the skewness, measure the bias, or asymmetry, of the distribution of data, while the kurtosis measures the degree of peakedness of the distribution of data around the average. The skewness is zero for symmetric data. The specification of these higher moments can be input into a cost function in exactly the same manner as the average. This was not incorporated in the present work.

9.3.2 Results for spherical scatterers

Figure 9.7 indicates that, for a given value of $\hat{n}$, Π depends in a mildly nonlinear manner on the particulate length scale ($\mathcal{L}$). Furthermore, there is a distinct minimum value of $\mathcal{L}$ to just block all of the incoming rays. A typical visualization for a simulation of the ray propagation is given in Figure 9.6. Clearly, the point where $\Pi = 0$, for each curve, represents the length scale that is just large enough to allow no rays to penetrate the system. For a given relative refractive index ratio, length scales larger than a critical value force more of the rays to be scattered backward. Table 9.1 indicates the estimated values for the length scale and

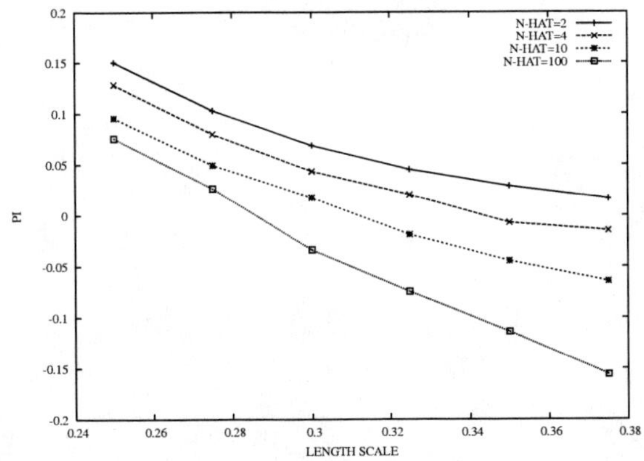

Figure 9.7. *The variation of* Π *as a function of* $\mathcal{L}$ *(Zohdi* [218]*).*

Table 9.1. *The estimated volume fractions needed for no complete penetration of incident electromagnetic energy,* $\Pi = 0$.

$\hat{n}$	$\mathcal{L}$	$v_p = \frac{4\pi\mathcal{L}^3}{3}$
2	0.4200	0.3107
4	0.3430	0.1692
10	0.3125	0.1278
100	0.2850	0.0969

the corresponding volume fraction needed to achieve no penetration of the electromagnetic rays, i.e., $\Pi = 0$. Clearly, at some point there are diminishing returns to increasing the volume fraction for a fixed refractive index ratio ($\hat{n}$). A least-squares curve fit indicates the following relationships between $\mathcal{L}$ and $\hat{n}$, as well as between the volume fraction v_p and $\hat{n}$, for $\Pi = 0$ to be achieved:

$$\mathcal{L} = 0.4090\hat{n}^{-0.0867} \quad \text{or} \quad v_p = 0.2869\hat{n}^{-0.2607}. \tag{9.42}$$

Qualitatively speaking, these results suggest the intuitive trend that if one has more reflective particles, one needs fewer of them to block (in a vectorially averaged sense) incoming rays, and vice versa.

To further understand this behavior, consider a single reflecting scatterer, with incident rays as shown in Figure 9.8. All rays at an incident angle between $\frac{\pi}{2}$ and $\frac{\pi}{4}$ are reflected with some positive y-component, i.e., "backward" (back scatter). However, between $\frac{\pi}{4}$ and 0, the rays are scattered with a negative y-component, i.e., forward. Since the reflectance is the ratio of the amount of reflected energy (irradiance) to the incident energy, it is appropriate to consider the integrated reflectance over a quarter of a single scatterer, which indicates the total fraction of the irradiance reflected:

$$\mathcal{I} \stackrel{\text{def}}{=} \frac{1}{\frac{\pi}{2}} \int_0^{\frac{\pi}{2}} R \, d\theta, \tag{9.43}$$

whose variation with $\hat{n}$ is shown in Figure 9.9. In the range tested of $2 \leq \hat{n} \leq 100$, the

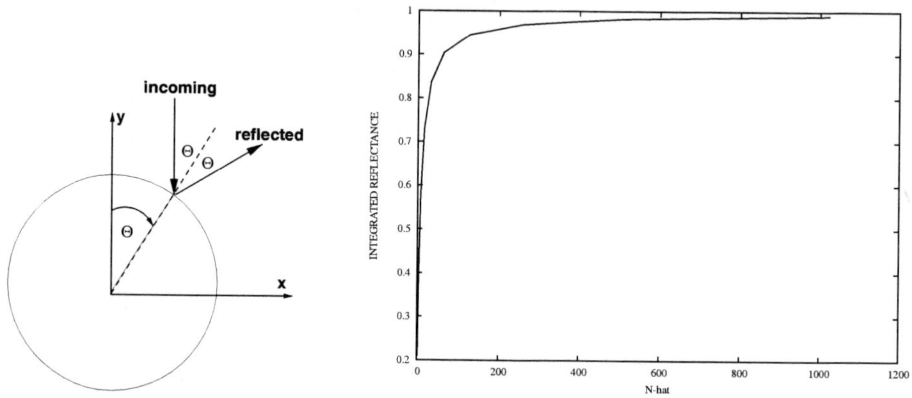

Figure 9.8. *Left, a single scatterer. Right, the integrated reflectance ($\mathcal{I}$) over a quarter of a single scatterer, which indicates the total fraction of the irradiance reflected (Zohdi* [219]*).*

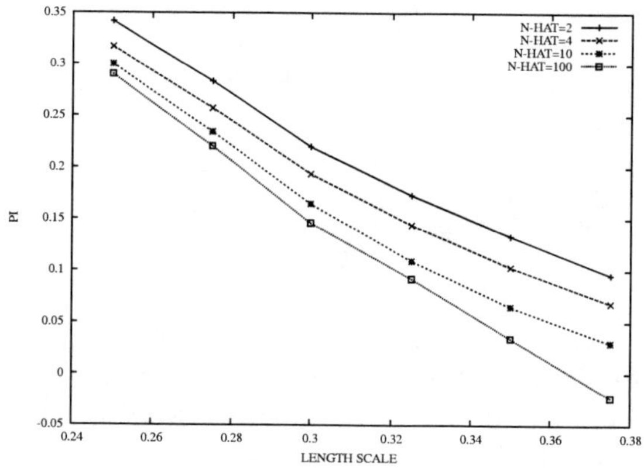

Figure 9.9. *(Oblate) Ellipsoids of aspect ratio* 4:1*: The variation of* Π *as a function of* $\mathcal{L}$*. The volume fraction is given by* $v_p = \frac{\pi \mathcal{L}^3}{4}$ *(Zohdi* [219]*).*

amount of energy reflected is a mildly nonlinear (quasi-linear) function of $\hat{n}$ for a single scatterer, and thus it is not surprising that it is the same for an aggregate.

9.3.3 Shape effects: Ellipsoidal geometries

One can consider a more detailed description of the scatterers, where we characterize the shape of the particles by the equation for an ellipsoid:[62]

$$F \stackrel{\text{def}}{=} \left(\frac{x - x_o}{r_1}\right)^2 + \left(\frac{y - y_o}{r_2}\right)^2 + \left(\frac{z - z_o}{r_3}\right)^2 = 1. \tag{9.44}$$

[62]The outward surface normals needed during the scattering process are relatively easy to characterize by writing $\boldsymbol{n} = \frac{\nabla F}{||\nabla F||}$. The orientation of the particles, usually random, can be controlled via rotational coordinate transformations.

As an example, consider oblate spheroids with an aspect ratio of $AR = \frac{r_1}{r_2} = \frac{r_1}{r_3} = 0.25$. As shown in Figure 9.9, the intuitive increase in volume fraction leads to an increase in overall reflectivity. The reason for this is that the volume fractions are so low, due to the fact that the particles are oblate, that the point of diminishing returns ($\Pi = 0$) is not met with the same length scale range as tested for the spheres. The volume fraction, for oblate spheroids given by $AR \leq 1$, is

$$v_p = \frac{4AR\pi\mathcal{L}^3}{3}, \tag{9.45}$$

where the largest radius (r_2 or r_3) is used to calculate $\mathcal{L}$. The volume fraction of a system containing oblate ellipsoidal particles, for example, with $AR = 0.25$, is much lower (one-sixteenth) than that of a system containing spheres with the same length scale parameter $\mathcal{L}$. As seen in Figure 9.9, at relatively high volume fractions ($\mathcal{L} = 0.375$), with the highest (idealized, mirror-like) reflectivity tested ($\hat{n} = 100$), the effect of "diminishing returns" begins, as it had for the spherical case. Clearly, it appears to be an effect that requires relatively high volume fractions to block the incoming rays, and consequently the effects of shape appear minimal for overall scattering.

Remark. Recently, a computational framework to rapidly simulate the light-scattering response of multiple red blood cells (RBCs), based upon ray-tracing, was developed in Zohdi and Kuypers [223]. Because the wavelength of visible light (roughly 3.8×10^{-7} m $\leq \lambda \leq 7.8 \times 10^{-7}$ m) is approximately at least an order of magnitude smaller than the diameter of a typical RBC scatterer ($d \approx 8 \times 10^{-6}$ m), geometric ray-tracing theory is applicable and can be used to quickly ascertain the amount of optical energy, characterized by the Poynting vector, that is reflected and absorbed by multiple RBCs. Three-dimensional examples were given to illustrate the approach, and the results compared quite closely to experiments on blood samples conducted at the Children's Hospital Oakland Research Institute (CHORI). See Appendix B for more details.

9.4 Discussion

For the disordered particulate systems considered, as the volume fraction of the scattering particles increases, as one would expect, less incident energy penetrates the aggregate particulate system. Above this critical volume fraction, more rays are scattered backward. However, the volume fraction at which the point of no penetration occurs depends in a quasi-linear fashion upon the ratio of the refractive indices of the particle and surrounding medium.

The similarity of electromagnetic scattering to acoustical scattering, governing sound disturbances that travels in inviscid media, is notable. Of course, the scales at which ray theory can be applied are much different because sound wavelengths are much larger than the wavelengths of light. The reflection of a plane harmonic pressure wave energy at an interface is given by[63]

$$R = \left(\frac{P_r}{P_i}\right)^2 = \left(\frac{\hat{A}\cos\theta_i - \cos\theta_t}{\hat{A}\cos\theta_i + \cos\theta_t}\right)^2, \tag{9.46}$$

where P_i is the incident pressure ray, P_r is the reflected pressure ray, $\hat{A} \stackrel{\text{def}}{=} \frac{\rho_t c_t}{\rho_i c_i}$, ρ_t is the medium the ray encounters (transmitted), c_t is the corresponding sound speed in that

[63]This relation is derived in Appendix B.

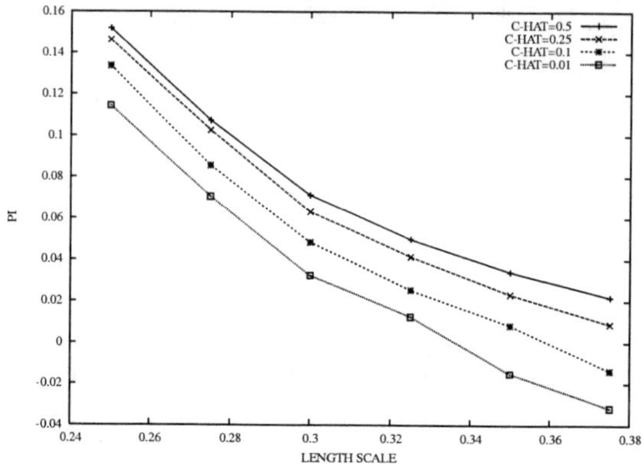

Figure 9.10. *Results for acoustical scattering ($\hat{c} = 1/\tilde{c}$) (Zohdi* [219]*).*

medium, ρ_i is the medium in which the ray was traveling (incident), and c_i is the corresponding sound speed in that medium. Clearly, the analysis of the aggregates can be performed for acoustical scattering in essentially the same way as for the optical problem. For example, for the same model problem as for the optical scenario (400 rays, 1000 scatterers), however, with the geometry and velocity appropriately scaled,[64] the results are shown in Figure 9.10 for varying $\hat{c} = \frac{c_t}{c_i} = 1/\tilde{c}$. The results for the acoustical analogy are quite similar to those for optics. See Appendix B for more details.

As mentioned earlier, for most materials the magnetic permeability is virtually the same, with exceptions being concentrated magnetite, pyrrhotite, and titanomagnetite (see Telford et al. [192] and Nye [153]). Clearly, with many new industrial materials being developed, possibly having nonstandard magnetic permeabilities ($\hat{\mu} \neq 1$), such effects may become more important to consider. Generally, from studying Equation (9.36), as $\hat{\mu} \to \infty$, $R \to 1$. In other words, as the relative magnetic permeability increases, the reflectance increases. More remarks are given in Appendix B.

Obviously, when more microstructural features are considered, for example, topological and thermal variables, parameter studies become quite involved. In order to eliminate a trial and error approach to determining the characteristics of the types of particles that would be needed to achieve a certain level of scattering, in Zohdi [218] an automated computational inverse solution technique has recently been developed to ascertain particle combinations that deliver prespecified electromagnetic scattering, thermal responses, and radiative (infrared) emission, employing genetic algorithms in combination with implicit staggering solution schemes, based upon approaches found in Zohdi [212]–[218]. This is discussed next.

9.5 Thermal coupling

The characterization of particulate systems, flowing or static, must usually be conducted in a nonevasive manner. Thus, experimentally speaking, light-scattering behavior can be a key

[64]Typical sound wavelengths are in the range of $0.01 \text{ m} \leq \lambda \leq 30 \text{ m}$, with wavespeeds in the range of $300 \text{ m/s} \leq c \leq 1500 \text{ m/s}$, thus leading to wavelengths, $f = c/\lambda$, with ranges on the order of $10 \text{ 1/s} \leq f \leq 150000 \text{ 1/s}$. Therefore, the scatterers must be much larger than scatterers in applications involving ray-tracing in optics.

indicator of the character of the flow. Experimentally speaking, thermal behavior can be a key indicator of the dynamical character of particulate flows. For example, in Chung et al. [45] and Shin et al. [177], techniques for measuring flow characteristics based upon infrared thermal velocimetry (ITV) in fluidic microelectromechanical systems (MEMS) have been developed. In such approaches, infrared lasers are used to generate a short heating pulse in a flowing liquid, and an infrared camera records the radiative images from the heated flowing liquid. The flow properties are obtained from consecutive radiative images. This approach is robust enough to measure particulate flows as well. In such approaches, a heater generates a short thermal pulse, and a thermal sensor detects the arrival downstream. This motivates the investigation of the coupling between optical scattering (electromagnetic energy propagation) and thermal coupling effects for particulate suspensions.

As before, it is assumed that the scattering particles are small enough to consider that the temperature fields are uniform in the particles.[65] We consider an energy balance, governing the interconversions of mechanical, thermal, and chemical energy in a system, dictated by the first law of thermodynamics. Accordingly, we require the time rate of change of the sum of the kinetic energy ($\mathcal{K}$) and stored energy ($\mathcal{S}$) to be equal to the sum of the work rate (power, $\mathcal{P}$) and the net heat supplied ($\mathcal{H}$):

$$\frac{d}{dt}(\mathcal{K}+\mathcal{S}) = \mathcal{P} + \mathcal{H}, \tag{9.47}$$

where the stored energy comprises a thermal part, $\mathcal{S}(t) = mC\theta(t)$, where C is the heat capacity per unit mass, and, consistent with our assumptions that the particles deform negligibly during the process, *a negligible mechanical stored energy portion.* The kinetic energy is $\mathcal{K}(t) = \frac{1}{2}m\boldsymbol{v}(t)\cdot\boldsymbol{v}(t)$. The mechanical power term is due to the total forces ($\boldsymbol{\Psi}^{tot}$) acting on a particle, namely,

$$\mathcal{P} = \frac{d\mathcal{W}}{dt} = \boldsymbol{\Psi}^{tot}\cdot\boldsymbol{v}. \tag{9.48}$$

Also, because $\frac{d\mathcal{K}}{dt} = m\dot{\boldsymbol{v}}\cdot\boldsymbol{v}(t)$, and we have a balance of momentum $m\dot{\boldsymbol{v}}\cdot\boldsymbol{v} = \boldsymbol{\Psi}^{tot}\cdot\boldsymbol{v}$, thus $\frac{d\mathcal{K}}{dt} = \frac{d\mathcal{W}}{dt} = \mathcal{P}$, leading to $\frac{d\mathcal{S}}{dt} = \mathcal{H}$. The primary source of heat is due to the incident rays. The energy input from the reflection of a ray is defined as

$$\Delta\mathcal{H}^{rays} \stackrel{\text{def}}{=} \int_t^{t+\Delta t} \mathcal{H}^{rays}\,dt \approx (I_i - I_r)a_r\Delta t = (1-R)I_i a_r \Delta t. \tag{9.49}$$

After an incident ray is reflected, it is assumed that a process of heat transfer occurs (Figure 9.11). It is assumed that the temperature fields are uniform within the particles; thus, conduction within the particles is negligible. We remark that the validity of using a lumped thermal model, i.e., ignoring temperature gradients and assuming a uniform temperature within a particle, is dictated by the magnitude of the Biot number. A small Biot number indicates that such an approximation is reasonable. The Biot number for spheres scales with the ratio of particle volume (V) to particle surface area (a_s), $\frac{V}{a_s} = \frac{b}{3}$, which indicates that a uniform temperature distribution is appropriate, since the particles, by definition, are small.

[65]Thus, the gradient of the temperature within the particle is zero, i.e., $\nabla\theta = \mathbf{0}$. Therefore, a Fourier-type law for the heat flux will register a zero value, $\boldsymbol{q} = -\mathbb{K}\cdot\nabla\theta = \mathbf{0}$. Furthermore, we assume that the space between the particles, i.e., the "ether," plays no role in the heat transfer process.

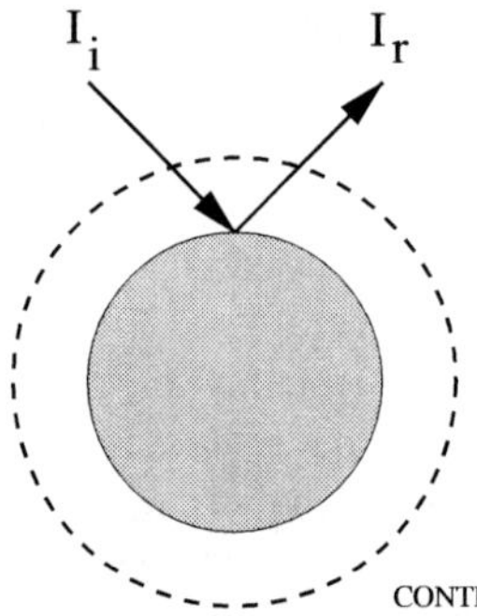

Figure 9.11. *Control volume for heat transfer (Zohdi* [218]*).*

The first law reads

$$\frac{d(K+S)}{dt} = m\dot{\boldsymbol{v}}\cdot\boldsymbol{v} + mC\dot{\theta} = \underbrace{\boldsymbol{\Psi}^{tot}\cdot\boldsymbol{v}}_{mechanical\ power} - \underbrace{h_c a_s(\theta-\theta_o)}_{convective\ heating} - \underbrace{\mathcal{B}a_s\varepsilon(\theta^4-\theta_s^4)}_{thermal\ radiation} + \underbrace{\mathcal{H}^{rays}}_{sources}, \tag{9.50}$$

where θ_o is the temperature of the ambient gas; θ_s is the temperature of the far-field surface (for example, a container surrounding the flow) with which radiative exchange is made; $\mathcal{B} = 5.67 \times 10^{-8}\frac{\mathrm{W}}{\mathrm{m^2\cdot K}}$ is the Stefan–Boltzmann constant; $0 \le \varepsilon \le 1$ is the emissivity, which indicates how efficiently the surface radiates energy compared to a black-body (an ideal emitter); $0 \le h_c$ is the heating due to convection (Newton's law of cooling) into the dilute gas; and a_s is the surface area of a particle. *It is assumed that the thermal radiation exchange between the particles is negligible.* For the applications considered here, typically, h_c is quite small and plays a small role in the heat transfer processes. From a balance of momentum we have $m\dot{\boldsymbol{v}}\cdot\boldsymbol{v} = \boldsymbol{\Psi}^{tot}\cdot\boldsymbol{v}$ and Equation (9.49) becomes

$$mC\dot{\theta} = -h_c a_s(\theta-\theta_o) - \mathcal{B}a_s\varepsilon(\theta^4-\theta_s^4) + \mathcal{H}^{rays}. \tag{9.51}$$

Therefore, after temporal integration with a finite difference time step of Δt, we have

$$\theta(t+\Delta t) = \frac{1}{mC + h_c a_s \Delta t}\left(mC\theta(t) - \Delta t\mathcal{B}a_s\varepsilon\left(\theta^4(t+\Delta t)-\theta_s^4\right) + \Delta t h_c a_s\theta_o + \Delta\mathcal{H}^{rays}\right). \tag{9.52}$$

This implicit nonlinear equation for θ, *for each particle*, is added into the ray-tracing algorithm in the next section.

9.6 Solution procedure

We now develop a staggering scheme by extending an approach found in Zohdi [208]–[210], [212], and [213]. After time discretization of the stored energy term in the equations of thermal equilibrium for a particle,

$$mC\dot{\theta}_i^{L+1} \approx mC\frac{\theta_i^{L+1}-\theta_i^L}{\Delta t}, \tag{9.53}$$

(⋆) COMPUTE RAY ORIENTATIONS AFTER REFLECTION (FRESNEL RELATIONS);

COMPUTE ABSORPTION CONTRIBUTIONS TO THE PARTICLES: $\Delta\mathcal{H}^{rays}$;

COMPUTE PARTICLE TEMP. (RECURSIVELY, $K = 1, 2, \ldots$ UNTIL CONVERGENCE):

$$\theta^{L+1,K} = \frac{1}{mC + h_c a_s \Delta t}\left(mC\theta^L - \Delta t \mathcal{B} a_s \varepsilon\left((\theta^{L+1,K-1})^4 - \theta_s^4\right) + \Delta t h_c a_s \theta_o + \Delta\mathcal{H}^{rays}\right);$$

INCREMENT ALL RAY POSITIONS: $\boldsymbol{r}_i(t + \Delta t) = \boldsymbol{r}_i(t) + \Delta t \boldsymbol{v}_i(t)$;

GO TO (⋆) AND REPEAT $(t = t + \Delta t)$.

ALGORITHM 9.2

where, for brevity, we write $\theta_i{}^{L+1} \stackrel{\text{def}}{=} \theta_i(t + \Delta t)$, $\theta_i{}^L \stackrel{\text{def}}{=} \theta_i(t)$, etc., we arrive at the abstract form, for the entire system, of $\mathcal{A}(\theta_i^{L+1}) = \mathcal{F}$. It is convenient to write

$$\mathcal{A}(\theta_i^{L+1}) - \mathcal{F} = \mathcal{G}(\theta_i^{L+1}) - \theta_i^{L+1} + \mathcal{R} = 0, \tag{9.54}$$

where $\mathcal{R}$ is a remainder term that does not depend on the solution, i.e., $\mathcal{R} \neq \mathcal{R}(\theta_i^{L+1})$. A straightforward iterative scheme can be written as

$$\theta_i^{L+1,K} = \mathcal{G}(\theta_i^{L+1,K-1}) + \mathcal{R}, \tag{9.55}$$

where $K = 1, 2, 3, \ldots$ is the index of iteration within time step $L + 1$. The convergence of such a scheme depends on the behavior of $\mathcal{G}$. Namely, a sufficient condition for convergence is that $\mathcal{G}$ be a contraction mapping for all $\theta_i^{L+1,K}$, $K = 1, 2, 3, \ldots$. In order to investigate this further, we define the error as $\Delta\theta^{L+1,K} = \theta_i^{L+1,K} - \theta_i^{L+1}$. A necessary restriction for convergence is iterative self-consistency, i.e., the exact solution must be represented by the scheme $\mathcal{G}(\theta_i^{L+1}) + \mathcal{R} = \theta_i^{L+1}$. Enforcing this restriction, a sufficient condition for convergence is the existence of a contraction mapping of the form

$$||\Delta\theta^{L+1,K}|| = ||\theta_i^{L+1,K} - \theta_i^{L+1}|| = ||\mathcal{G}(\theta_i^{L+1,K-1}) - \mathcal{G}(\theta_i^{L+1})|| \leq \eta^{L+1,K}||\theta_i^{L+1,K-1} - \theta_i^{L+1}||, \tag{9.56}$$

where, if $\eta^{L+1,K} < 1$ for each iteration K, then $\Delta\theta^{L+1,K} \to 0$ for any arbitrary starting value $\theta_i^{L+1,K=0}$ as $K \to \infty$. The type of contraction condition discussed is sufficient, but not necessary, for convergence. Typically, the time step sizes for ray-tracing are far smaller than needed; thus, the approach converges quickly. More specifically, $\mathcal{G}$'s behavior is controlled by $\frac{\Delta t \mathcal{B} a_s \varepsilon}{mC + h_c a_s \Delta t}$, which is quite small. Thus, a fixed-point iterative scheme, such as the one introduced, converges rapidly. This iterative procedure is embedded into the overall ray-tracing scheme. For the overall algorithm (starting at $t = 0$ and ending at $t = T$), see Algorithm 9.2.

In order to capture all of the internal reflections that occur when rays enter the particulate systems, the time step size Δt is dictated by the size of the particles. A somewhat ad hoc approach is to scale the time step size according to $\Delta t = \xi b$, where b is the radius of the particles and typically $0.05 \leq \xi \leq 0.1$.

9.7 Inverse problems/parameter identification

An important aspect of any model is the inverse problem of identifying parameters that force the system behavior to match a target response and may stem from an experimental observation or a design specification. In the ideal case, one would like to determine combinations of scattering parameters that produce certain aggregate effects, via numerical simulations, in order to minimize time-consuming laboratory tests. The primary quantity of interest in this work is the percentage of lost irradiance by a beam in a selected direction over the time interval of $(0, T)$. As in the previous examples, this is characterized by the inner product of the Poynting vector and a selected direction ($\boldsymbol{d}$):

$$\mathcal{Z}(0, T) \stackrel{\text{def}}{=} \frac{\sum_{i=1}^{N_r} (\boldsymbol{S}(t = 0) - \boldsymbol{S}(t = T)) \cdot \boldsymbol{d}}{\sum_{i=1}^{N_r} \boldsymbol{S}_i(t = 0) \cdot \boldsymbol{d}}, \tag{9.57}$$

where $\mathcal{Z}$ can be considered the amount of energy "blocked" (in a vectorially averaged sense) from propagating in the $\boldsymbol{d}$ direction. Now consider a cost function comparing the loss to the specified blocked amount:

$$\Pi \stackrel{\text{def}}{=} \left| \frac{\mathcal{Z}(0, T) - \mathcal{Z}^*}{\mathcal{Z}^*} \right|, \tag{9.58}$$

where the total simulation time is T and where $\mathcal{Z}^*$ is a target blocked value. One can augment this by also monitoring the average temperature of the scattering particles during the time interval,

$$\Theta(0, T) \stackrel{\text{def}}{=} \frac{1}{N_p T} \int_0^T \sum_{i=1}^{N_p} \theta_i(t)\, dt, \tag{9.59}$$

as well as the average emitted thermal radiation of the scatterers during the time interval,

$$\Gamma(0, T) \stackrel{\text{def}}{=} \frac{1}{N_p T} \int_0^T \sum_{i=1}^{N_p} \mathcal{B} a_{si} \varepsilon_i (\theta_i^4(t) - \theta_s^4)\, dt, \tag{9.60}$$

to yield the composite cost function

$$\Pi(w_1, w_2, w_3) \stackrel{\text{def}}{=} \frac{1}{\sum_{j=1}^{3} w_j} \left(w_1 \left| \frac{\mathcal{Z}(0, T) - \mathcal{Z}^*}{\mathcal{Z}^*} \right| + w_2 \left| \frac{\Theta(0, T) - \Theta^*}{\Theta^*} \right| + w_3 \left| \frac{\Gamma(0, T) - \Gamma^*}{\Gamma^*} \right| \right), \tag{9.61}$$

where Θ^* and Γ^* are specified values. Typically, for the class of problems considered in this work, formulations such as in Equation (9.61) depend in a nonconvex and nondifferentiable manner on the system parameters. With respect to the minimization of Equation (9.61), classical gradient-based deterministic optimization techniques are not robust due to difficulties with objective function nonconvexity and nondifferentiability. Classical gradient-based algorithms are likely to converge only toward a local minimum of the objective function if an accurate initial guess for the global minimum is not provided. Also, usually it is extremely difficult to construct an initial guess that lies within the (global) convergence radius of a gradient-based method. These difficulties can be circumvented by using a certain class of nonderivative search methods, i.e., genetic algorithms, before applying gradient-based schemes. Genetic algorithms are search methods based on the principles of natural selection, employing concepts of species evolution such as reproduction, mutation, and crossover. Implementation typically involves a randomly generated population of fixed-length elemental strings, "genetic information," each of which represents a specific choice of system parameters. The population of individuals undergoes "mating sequences" and other biologically inspired events in order to find promising regions of the search space. There are a variety of such methods, employing concepts of species evolution such as reproduction, mutation, and crossover. Such methods primarily stem from the work of John Holland (Holland [94]). For reviews of such methods, see, for example, Goldberg [77], Davis [50], Onwubiko [155], Kennedy and Eberhart [120], Lagaros et al. [129], Papadrakakis et al. [156]–[159] and Goldberg and Deb [78].

Remark. To compute the fitness of a parameter set, one must go through the procedure in Algorithm 9.2, requiring a full-scale simulation. It is important to scale the system variables, for example, to be positive numbers and of comparable magnitude, in order to avoid dealing with large variations in the parameter vector components. Typically, for particulate flows with a finite number of particles, there will be slight variations in the performance for different random starting configurations. In order to stabilize the objective function's value with respect to the randomness of the flow starting configuration, for a given parameter selection ($\mathbf{\Lambda}$), a regularization procedure is applied within the genetic algorithm, whereby the performances of a series of different random starting configurations are averaged until the (ensemble) average converges, i.e., until the following condition is met:

$$\left| \frac{1}{Z+1} \sum_{i=1}^{Z+1} \Pi^{(i)}(\mathbf{\Lambda}^I) - \frac{1}{Z} \sum_{i=1}^{Z} \Pi^{(i)}(\mathbf{\Lambda}^I) \right| \leq TOL \left| \frac{1}{Z+1} \sum_{i=1}^{Z+1} \Pi^{(i)}(\mathbf{\Lambda}^I) \right|,$$

where index i indicates a different starting random configuration ($i = 1, 2, \ldots, Z$) that has been generated and Z indicates the total number of configurations tested. In order to implement this in the genetic algorithm, in Step 2, one simply replaces *compute* with *ensemble compute*, which requires a further inner loop to test the performance of multiple starting configurations. Similar ideas have been applied to other types of randomly dispersed particulate media in Zohdi [208]–[213]. *Clearly, such a procedure is not necessary when the scatterers are periodically arranged.*

Remark. As before, the classical random sequential addition algorithm was used to place nonoverlapping particles into the domain of interest (Widom [200]). This algorithm was adequate for the volume fraction ranges of interest (under 30%).

9.8 Parametrization and a genetic algorithm

We considered a group of N_p randomly positioned particles, of equal size, in a cube of normalized dimensions, $D \times D \times D$, with D normalized to unity. The particle size and volume fraction were determined by a particle/sample size ratio, which was defined via a subvolume size $V \stackrel{\text{def}}{=} \frac{D \times D \times D}{N_p}$, where N_p was the number of particles in the entire cube (Figure 9.12). The ratio between the radius (b) and the subvolume was denoted by

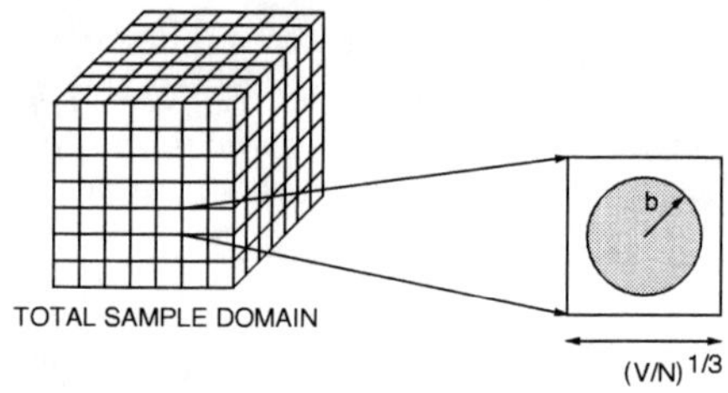

Figure 9.12. *Definition of a particle length scale (Zohdi* [218]*).*

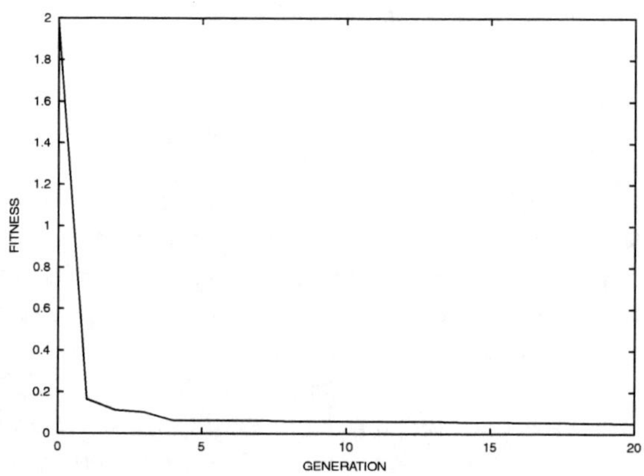

Figure 9.13. *The best parameter set's objective function values for successive generations. Note: The first data point in the optimization corresponds to the objective function's value for mean parameter values of upper and lower bounds of the search intervals (Zohdi* [218]*).*

$\mathcal{L} \stackrel{\text{def}}{=} \frac{b}{V^{\frac{1}{3}}}$. The volume fraction occupied by the particles was $v_p \stackrel{\text{def}}{=} \frac{4\pi\mathcal{L}^3}{3}$. Thus, the total volume occupied by the particles, denoted by ν, can be written as $\nu = v_p N_p V$. We used $N_p = 1000$ particles and $N_r = 400$ rays, arranged in a square 20×20 pattern (Figure 9.14). This system provided stable results, i.e., increasing the number of rays and/or the number of particles beyond these levels resulted in negligibly different overall system responses. The free parameters in the inverse problem were as follows:

- The particle length scale was $0 < \mathcal{L} \leq 0.35$.
- The relative refractive index ratio was $1 < \hat{n} \leq 10$.
- The particle emissivity was $0 \leq \varepsilon \leq 1$.
- The particle density, combined with the heat capacity, was $(\rho C)^- \leq (\rho C) \leq (\rho C)^+$, where $mC = \rho\frac{4}{3}\pi b^3 C$. C was held fixed at $C = 10^3$ N $\cdot$ m/°K and 10^3 kg/m^3 $=$ $\rho^- \leq \rho \leq \rho^+ = 2 \times 10^3$ kg/m^3.

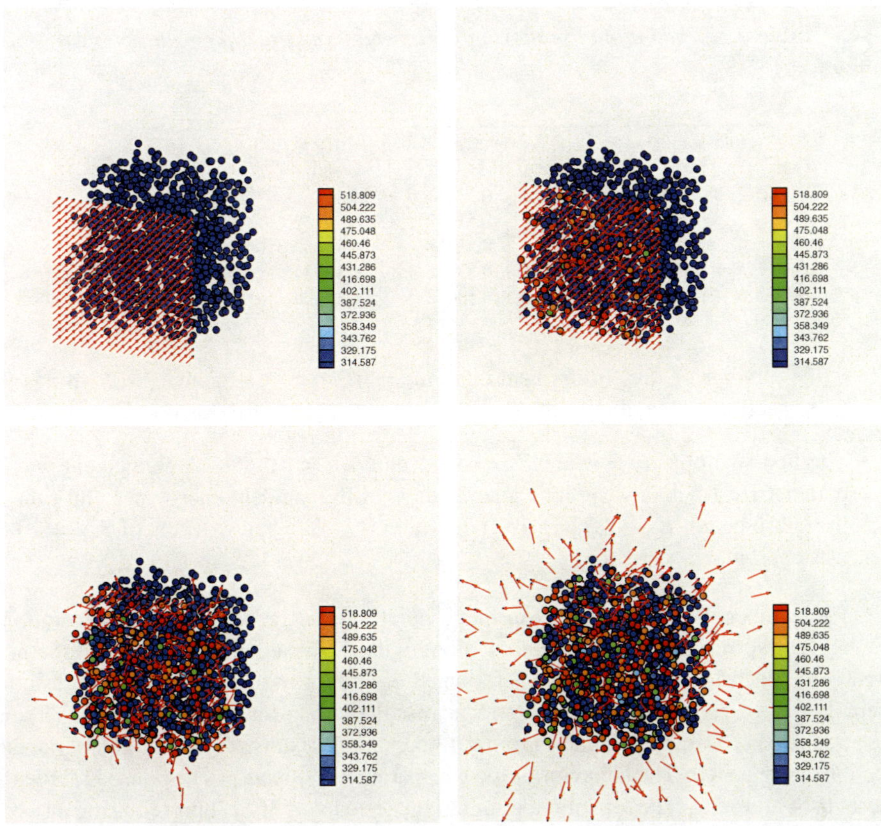

Figure 9.14. *Top to bottom and left to right, the progressive movement of rays making up a beam (for the best inverse parameter set vector (Table 9.2)). The colors of the particles indicate their temperature and the lengths of the vectors indicate the irradiance magnitude (Zohdi* [218]*).*

Thus, explicitly, the genetic string comprised the following parameters:

$$\mathbf{\Lambda} = (\mathcal{L}, \rho C, \epsilon, \hat{n}). \tag{9.62}$$

Other simulation parameters of importance are as follows:

- The dimensions of the sample were 10^{-3} m $\times 10^{-3}$ m $\times 10^{-3}$ m.
- The time scale was set to $\frac{3\times 10^{-3}\,\mathrm{m}}{c}$, where $c = 3 \times 10^8$ m/s is the speed of light.
- The initial velocity vector for all initially collinear rays making up the beam was $\boldsymbol{v} = (c, 0, 0)$.
- The irradiance beam parameter was set to $I = 10^{18}\ \mathrm{N \cdot m/(m^2 \cdot s)}$, where the irradiance for each ray was calculated as $I^{ray}(t=0)a_r \stackrel{\text{def}}{=} Ia_b/N_r$, where $N_r = 20 \times 20 = 400$

Table 9.2. *The optimal scattering parameters and the top six fitnesses with* $w_1 = w_2 = w_3 = 1$.

Rank	$\mathcal{L}$	$\hat{n}$	ε	$\rho \times 10^{-3}$ kg/m^3	Π
1	0.21480	5.82056	0.53687	0.15078	0.04968310
2	0.21481	5.91242	0.53741	0.15152	0.05126406
3	0.21482	5.89121	0.53637	0.15152	0.05166210
4	0.21482	5.83350	0.53636	0.15150	0.05232877
5	0.21477	6.23032	0.53748	0.16034	0.05236720
6	0.21481	5.81637	0.53672	0.15008	0.05260397

is the number of rays in the beam and $a_b = 10^{-3}$ m $\times 10^{-3}$ m $= 10^{-6}$ m^2 is the cross-sectional area of the beam.

- The first two objectives were $\mathcal{Z}^* = 0.75$ and $\Theta^* = 400°$ K. A convenient way to parametrize Γ^* is to write it as a percentage of the incident energy per unit time of the entire beam, $K^* I^{ray}(t = 0) \times N_r$, where $0 \leq K^* \leq 1$. A value of $K^* = 10^{-18}$ was chosen.

The number of genetic strings in the population was set to 20, for 20 generations, allowing 6 offspring of the top 6 parents, along with their parents, to proceed to the next generation. Therefore, after each generation, 8 entirely new genetic strings were introduced. Every 10 generations, the search was rescaled around the best parameter set, and the search restarted. Table 9.2 and Figure 9.13 depict the results. A total of 286 parameter selections were tested. The behavior of the best parameter selection's response is shown in Figures 9.14 and 9.15. The total number of strings tested was 3651, thus requiring an average of 12.765 strings per parameter selection for the ensemble averaging stabilization. After approximately 6 generations, the procedure stabilized. We again remark that gradient-based methods are sometimes useful for postprocessing solutions found with a genetic algorithm, if the objective function is sufficiently smooth in that region of the parameter space. This was not done in this work; however, the reader can consult the texts of Luenberger [142] and Gill et al. [76], or the survey in Papadrakakis et al. [160].

9.9 Summary

The presented work developed a ray-tracing algorithm that was combined with a stochastic genetic algorithm in order to treat coupled inverse optical scattering formulations, where physical parameters, such as particulate volume fractions, refractive indices, and thermal constants, were sought so that the overall response of a sample of randomly distributed particles, suspended in an ambient medium, would match desired coupled scattering, thermal, and infrared responses. Large-scale numerical simulations were presented to illustrate the overall procedure and to investigate aggregate ray dynamics corresponding to the flow of electromagnetic energy and the conversion of the absorbed energy into heat and infrared radiation through disordered particulate systems.

Such design methodologies may be helpful in designing optical coating materials comprising randomly dispersed particles suspended in a binding matrix. The matrix usually

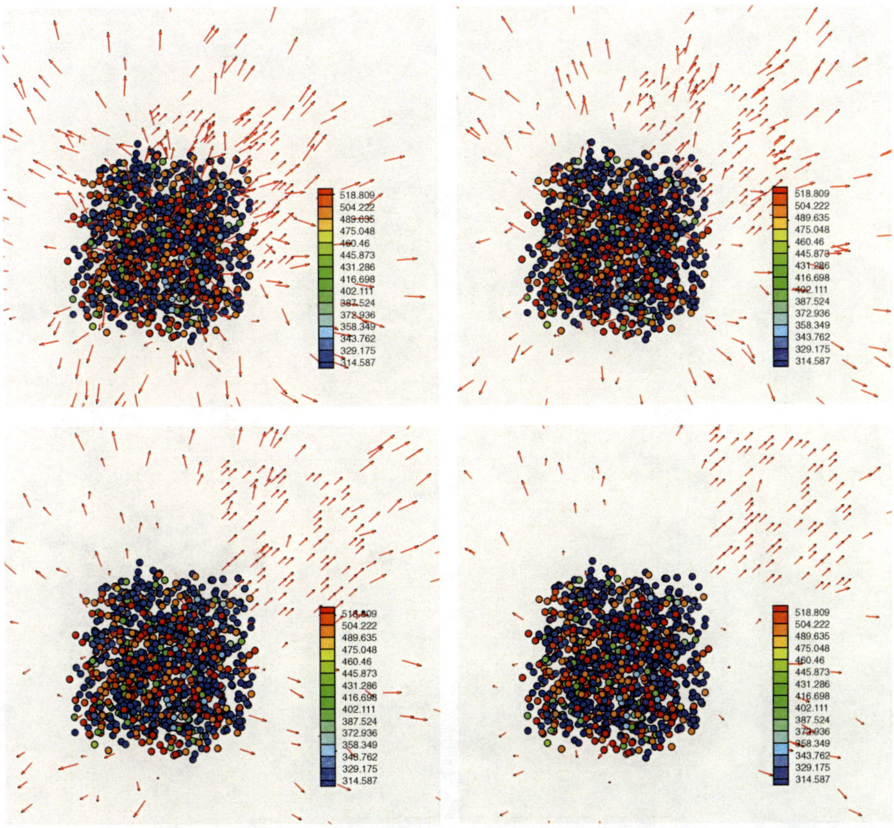

Figure 9.15. *Continuing Figure* 9.14, *top to bottom and left to right, the progressive movement of rays making up a beam (for the best inverse parameter set vector (Table* 9.2*)). The colors of the particles indicate their temperature and the lengths of the vectors indicate the irradiance magnitude (Zohdi* [218]*).*

has good adhesive and mechanical properties, while the particles are used as scattering units. Such coatings are relatively inexpensive to fabricate. The overall optical properties of such materials can be tailored by adjusting the volume fraction and refractive index of the particulate additives.

Accordingly, we can consider a more detailed description of the scatterers, where we characterize the shape of the particles by a generalized ellipsoidal equation:[66]

$$F \stackrel{\text{def}}{=} \left(\frac{|x - x_o|}{r_1}\right)^{s_1} + \left(\frac{|y - y_o|}{r_2}\right)^{s_2} + \left(\frac{|z - z_o|}{r_3}\right)^{s_3} = 1, \tag{9.63}$$

where the s's are exponents. The orientation of the particles, usually random, can be controlled via rotational coordinate transformations. Values of $s < 1$ produce nonconvex

[66]The outward surface normals, $\boldsymbol{n}$, needed during the scattering calculations, are relatively easy to characterize by writing $\boldsymbol{n} = \frac{\nabla F}{||\nabla F||}$.

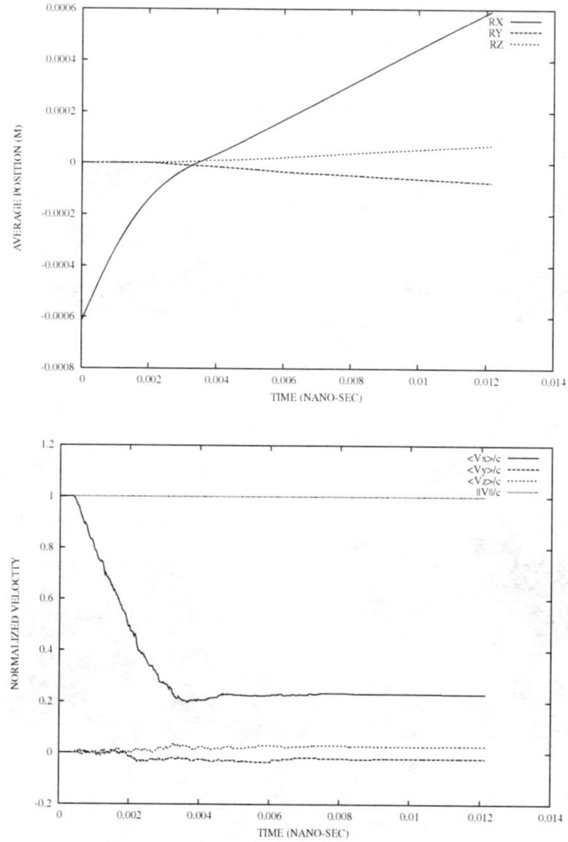

Figure 9.16. *Top, the components of the average position over time for the best parameter set. Bottom, the components of the average ray velocity and the Euclidean norm over time for the best parameter set. The normalized quantity* $||\boldsymbol{v}||/c = 1$ *serves as a type of computational "error check" (Zohdi* [218]*).*

shapes, while $s > 2$ values produce "block-like" shapes (three inverse parameters). Furthermore, we can introduce the particulate aspect ratio, defined by $AR \stackrel{\text{def}}{=} \frac{r_1}{r_2} = \frac{r_1}{r_3}$, where $r_2 = r_3$, $AR > 1$ for prolate geometries, and $AR < 1$ for oblate shapes (one variable). Therefore, including the variables introduced before, in the most general case we have a total of nine variables, $\boldsymbol{\Lambda} = (\mathcal{L}, \rho C, \epsilon, \hat{n}, \hat{\mu}, s_1, s_2, s_3, AR)$. We remark that if the particles' orientations are assumed aligned, then three more (angular orientation) parameters can be introduced, $(\theta_1, \theta_2, \theta_3)$. In fact, suspensions can become aligned, for example, along electrical field lines induced by external sources, or due to flow conditions. Thus, the search space grows to 12 parameters,[67] $\boldsymbol{\Lambda} = (\mathcal{L}, \rho C, \epsilon, \hat{n}, \hat{\mu}, s_1, s_2, s_3, AR, \theta_1, \theta_2, \theta_3)$.

[67] It is important to note that the control of the particle properties, volume fractions, orientations, etc., can be used to design hybrid thin films composed of particulate additives in a matrix binder.

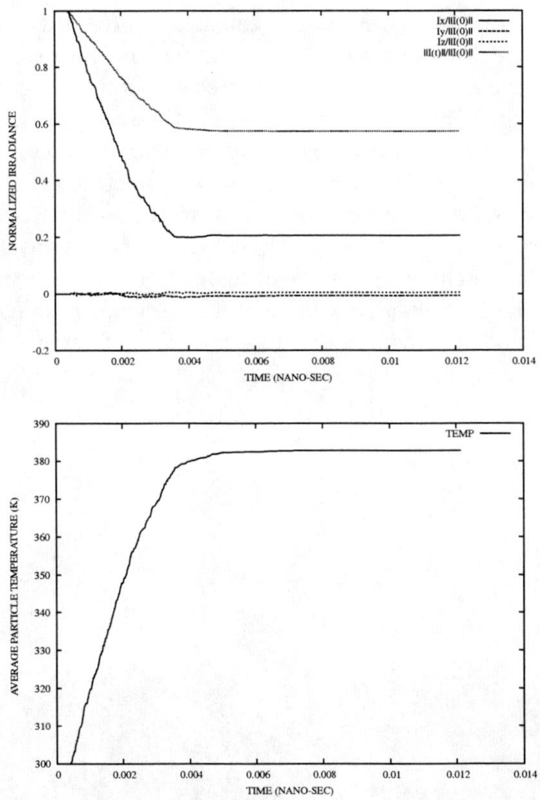

Figure 9.17. *Top, the components of the average ray irradiance and the Euclidean norm over time for the best parameter set. Bottom, the average temperature of the scatterers over time for the best parameter set (Zohdi [218]).*

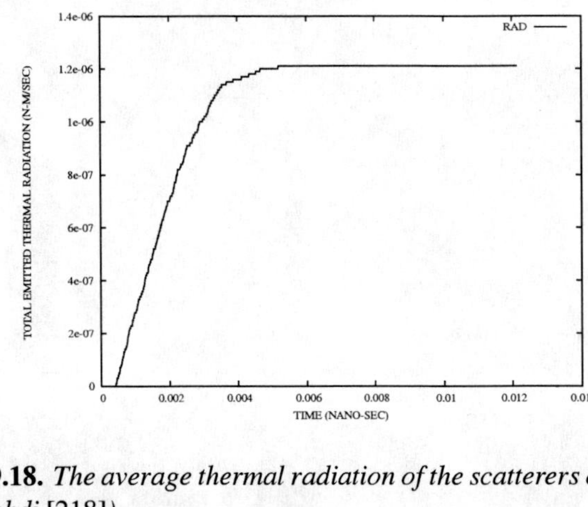

Figure 9.18. *The average thermal radiation of the scatterers over time for the best parameter set (Zohdi [218]).*

Finally, in addition to a more detailed characterization of the particle geometry, in some cases transparent particle materials, accounting for refractive and dispersive rays traveling through scatterers, can be important. Recall that the dispersion of a light ray is how, for example, white light, which is a mixture of all wavelengths of visible light, can be decomposed into its constituent wavelengths or colors when it passes from one medium into another. This phenomenon occurs because the index of refraction of a transparent medium is greater for light of shorter wavelengths. Thus, whenever light is refracted in passing from one medium to the next, the violet and blue light of shorter wavelengths is bent more than the orange and red light of longer wavelengths.[68] Thus, dispersive effects introduce a new level of complexity, primarily because of the refraction of different wavelengths of light, leading to a dramatic growth in the number of rays of varying intensities and color (wavelength). The inclusion of these effects is currently under investigation by the author.

[68]This is how a rainbow is formed.

Chapter 10

Closing remarks

This monograph provided a basic introduction to the subject of particulate flows. Clearly, a comprehensive survey of all the possible modeling and computational techniques cannot be undertaken in a work of this size. However, an extensive list of references has been provided. In particular, we note that a survey of fast computational methods, specifically efficient contact search techniques for the treatment of densely packed granular or particulate media, in the absence of near-field forces, can be found in the recent work of Pöschel and Schwager [167]. However, while such techniques are outside the scope of the present work, they are relatively easy to implement and are highly recommended to attain high-performance simulations for large numbers of particles, in particular when they are irregularly shaped.

Applications for the models developed include industrial processes such as chemical mechanical planarization (CMP), which involves using particles embedded in fluid (gas or liquid) to ablate small-scale surfaces flat. Such processes have become important for the success of many micro- and nanotechnologies, such as integrated circuit fabrication. However, the process is still one of trial and error. During the last decade, understanding of the basic mechanisms involved in this process has initiated research efforts in both industry and academia. For a review of CMP practice and applications, see Luo and Dornfeld [143]–[146]. It is clear that for the process to become viable and efficient, the underlying physics must be modeled in a detailed, nonphenomenological manner. Ultimately, the ability to perform rapid computational simulation of particle dynamics raises the possibility to *optimize CMP-related parameters*, such as particle sizes, distributions, densities, and grinding-pad surfaces, for a given application.

In the natural sciences, the study of particle-laden dust clouds, stemming from ejecta (nickel, magnesium, and iron) from comets and asteroids, is becoming increasingly important. A prominent example is the famous Tempel–Tuttle comet, which passes through the solar system every 33 years. When the ejecta from this comet intersect the orbits of satellites, a number of difficulties can occur. Due to the increasingly rapid commercialization of near-Earth space and the presence of thousands of satellites, space-dust/satellite interaction problems are becoming of greater concern. Most larger objects, down to about the 0.1-m level, are tracked in low-Earth orbit. However, it is simply infeasible to track smaller-sized dust.[69] For example, so-called Leonids, millimeter-level clouds, so named because they appear to radiate from the head of the constellation of Leo the Lion, have been blamed for the malfunction of several satellites (Brown and Cooke [37]). There are many more such debris clouds, such as Draconids, Lyrids, Peresids, and Andromedids, which are named for the constellations from which they appear to emanate. Such debris may lead not only to mechanical damage to the satellites but also to instrumentation failure by disintegrating into charged particle-laden plasmas, which affect the sensitive electrical components on board. In another space-related area, dust clouds are also important in the formation of planetesimals, which are thought to be initiated by the agglomeration of dust particles. For more information see Benz [26], [27], Blum and Wurm [32], Dominik and Tielens [54], Chokshi et al. [43], Wurm et al. [204], Kokubu and Ida [127], [128], Mitchell and Frenklach [148], Grazier et al. [83], [84], Supulver and Lin [182], Tanga et al. [191], Cuzzi et al. [48], Weidenschilling and Cuzzi [198], Weidenschilling et al. [199], Beckwith et al. [20], Barge and Sommeria [14], Pollack et al. [166], Lissauer [138], Barranco et al. [15], and Barranco and Marcus [16], [17].

In closing, it is important to mention related particle-laden flow problems arising from the analysis of biological systems. Specifically, there are numerous applications in biomechanics where one step in an overall series of events is the collision and possible adhesion of small-scale particles, under the influence of near-fields. For example, in the study of atherosclerotic plaque growth, a predominant school of thought attributes the early stages of the disease to a relatively high concentration of microscale suspensions (low-density lipoprotein (LDL) particles) in blood.[70] Atherosclerotic plaque formation involves (a) adhesion of monocytes (essentially larger suspensions) to the endothelial surface, which is controlled by the adhesion molecules stimulated by the excess LDL, the oxygen content, and the intensity of the blood flow; (b) penetration of the monocytes into the intima and subsequent tissue inflammation; and (c) rupture of the plaque accompanied by some degree of thrombus formation or even subsequent occlusive thrombosis. For surveys, see Fuster [72], Shah [174], van der Wal and Becker [197], Chyu and Shah [46], and Libby [134], [135], Libby et al. [136], Libby and Aikawa [137], Richardson et al. [169], Loree et al. [141], and Davies et al. [51], among others. The mechanisms involved in the initial stages of the disease, in particular stage (a), have not been extensively studied, although some simple semi-analytical qualitative studies have been carried out recently in Zohdi et al. [220] and Zohdi [221], in particular focusing on particle adhesion to artery walls. Furthermore, particle-to-particle adhesion can play a significant role in the behavior of a thrombus, comprising agglomerations of particles, ejected by a plaque rupture. The behavior, in particular the fragmentation, of such a thrombus as it moves downstream is critical in determining the chances for stroke. For extensive analyses addressing modeling and numerical procedures, see Kaazempur-Mofrad and Ethier [113], Williamson et al. [202], Younis et al. [205], Kaazempur-Mofrad et al. [114], Kaazempur-Mofrad et al. [115], Chau et al. [41], Chan et al. [38], Dai et al. [49], Khalil et al. [121], Khalil et al. [122], Stroud et al. [180], [181], Berger and Jou [29], and Jou and Berger [112]. For experimentally oriented physiological flow studies of atherosclerotic carotid bifurcations and related systems, see Bale-Glickman et al. [12], [13]. Notably, Bale-Glickman et al. [12], [13] have

[69] Ground-based radar and optical and infrared sensors routinely track several thousand objects daily.

[70] Plaques with high risk of rupture are termed *vulnerable*.

constructed flow models that replicate the lumen of plaques excised intact from patients with severe atherosclerosis, which have shown that the complex internal geometry of the diseased artery, combined with the pulsatile input flows, gives exceedingly complex flow patterns. They have shown that the flows are highly three-dimensional and chaotic, with details varying from cycle to cycle. In particular, the vorticity and streamline maps confirm the highly complex and three-dimensional nature of the flow. Another biological process where particle interaction and aggregation is important is the formation of certain types of kidney stones, which start as an agglomeration "seed" of particulate materials, for example, combinations of calcium oxalate monohydrate, calcium oxalate dihydrate, uric acid, struvite, or cystine. For details, see Coleman and Saunders [47], Kim [124], Pittomvils et al. [165], Kahn et al. [116], Kahn and Hackett [117], [118], and Zohdi and Szeri [222]. Clearly, the number of applications in the biological sciences is enormous and growing. More general information on the theory and simulations found in this monograph can be found at `http://www.siam.org/books/cs04`.

Appendix A

Basic (continuum) fluid mechanics

The term "deformation" refers to a change in the shape of the continuum between a reference configuration and the current configuration. In the reference configuration, a representative particle of the continuum occupies a point $\mathbf{p}$ in space and has the position vector

$$\boldsymbol{X} = X_1\boldsymbol{e}_1 + X_2\boldsymbol{e}_2 + X_3\boldsymbol{e}_3,$$

where $\boldsymbol{e}_1, \boldsymbol{e}_2, \boldsymbol{e}_3$ is a Cartesian reference triad, and X_1, X_2, X_3 (with center $\mathbf{O}$) can be thought of as labels for a point. Sometimes the coordinates or labels (X_1, X_2, X_3, t) are called the referential coordinates. In the current configuration, the particle originally located at point $\mathbf{p}$ is located at point $\mathbf{p}'$ and can also be expressed in terms of another position vector $\boldsymbol{x}$ with the coordinates (x_1, x_2, x_3, t). These are called the current coordinates. It is obvious with this arrangement that the displacement is $\boldsymbol{u} = \boldsymbol{x} - \boldsymbol{X}$ for a point originally at $\boldsymbol{X}$ and with final coordinates $\boldsymbol{x}$.

When a continuum undergoes deformation (or flow), its points move along various paths in space. This motion may be expressed by

$$\boldsymbol{x}(X_1, X_2, X_3, t) = \boldsymbol{u}(X_1, X_2, X_3, t) + \boldsymbol{X}(X_1, X_2, X_3, t),$$

which gives the present location of a point at time t, written in terms of the labels X_1, X_2, X_3. The previous position vector may be interpreted as a mapping of the initial configuration onto the current configuration. In classical approaches, it is assumed that such a mapping is one-to-one and continuous, with continuous partial derivatives to whatever order is required. The description of motion or deformation expressed previously is known as the Lagrangian formulation. Alternatively, if the independent variables are the coordinates $\boldsymbol{x}$ and t, then $\boldsymbol{x}(x_1, x_2, x_3, t) = \boldsymbol{u}(x_1, x_2, x_3, t) + \boldsymbol{X}(x_1, x_2, x_3, t)$, and the formulation is called Eulerian.

A.1 Deformation of line elements

Partial differentiation of the displacement vector $\boldsymbol{u} = \boldsymbol{x} - \boldsymbol{X}$, with respect to $\boldsymbol{x}$ and $\boldsymbol{X}$, produces the displacement gradients

$$\nabla_X \boldsymbol{u} = \boldsymbol{F} - \mathbf{1} \quad \text{and} \quad \nabla_x \boldsymbol{u} = \mathbf{1} - \overline{\boldsymbol{F}}, \tag{A.1}$$

where

$$\nabla_x \boldsymbol{x} \stackrel{\text{def}}{=} \frac{\partial \boldsymbol{x}}{\partial \boldsymbol{X}} = \boldsymbol{F} \stackrel{\text{def}}{=} \begin{bmatrix} \frac{\partial x_1}{\partial X_1} & \frac{\partial x_1}{\partial X_2} & \frac{\partial x_1}{\partial X_3} \\ \frac{\partial x_2}{\partial X_1} & \frac{\partial x_2}{\partial X_2} & \frac{\partial x_2}{\partial X_3} \\ \frac{\partial x_3}{\partial X_1} & \frac{\partial x_3}{\partial X_2} & \frac{\partial x_3}{\partial X_3} \end{bmatrix} \tag{A.2}$$

and

$$\nabla_x \boldsymbol{X} \stackrel{\text{def}}{=} \frac{\partial \boldsymbol{X}}{\partial \boldsymbol{x}} = \overline{\boldsymbol{F}}, \tag{A.3}$$

with the components $F_{ik} = x_{i,k}$ and $\overline{F}_{ik} = X_{i,k}$. $\boldsymbol{F}$ is known as the material deformation gradient and $\overline{\boldsymbol{F}}$ is known as the spatial deformation gradient.

Remark. It should be clear that $d\boldsymbol{x}$ can be reinterpreted as the result of a mapping $\boldsymbol{F} \cdot d\boldsymbol{X} \to d\boldsymbol{x}$, or a change in configuration (reference to current), while $\overline{\boldsymbol{F}} \cdot d\boldsymbol{x} \to d\boldsymbol{X}$ maps the current to the reference system. For the deformations to be invertible, and physically realizable, $\overline{\boldsymbol{F}} \cdot (\boldsymbol{F} \cdot d\boldsymbol{X}) = d\boldsymbol{X}$ and $\boldsymbol{F} \cdot (\overline{\boldsymbol{F}} \cdot d\boldsymbol{x}) = d\boldsymbol{x}$. We note that $(\det \overline{\boldsymbol{F}})(\det \boldsymbol{F}) = 1$ and we have the obvious relation $\frac{\partial \boldsymbol{X}}{\partial \boldsymbol{x}} \cdot \frac{\partial \boldsymbol{x}}{\partial \boldsymbol{X}} = \overline{\boldsymbol{F}} \cdot \boldsymbol{F} = \mathbf{1}$. It should be clear that $\overline{\boldsymbol{F}} = \boldsymbol{F}^{-1}$.

A.2 The Jacobian of the deformation gradient

The Jacobian of the deformation gradient $\boldsymbol{F}$ is defined as

$$J \stackrel{\text{def}}{=} \det \boldsymbol{F} = \begin{vmatrix} \frac{\partial x_1}{\partial X_1} & \frac{\partial x_1}{\partial X_2} & \frac{\partial x_1}{\partial X_3} \\ \frac{\partial x_2}{\partial X_1} & \frac{\partial x_2}{\partial X_2} & \frac{\partial x_2}{\partial X_3} \\ \frac{\partial x_3}{\partial X_1} & \frac{\partial x_3}{\partial X_2} & \frac{\partial x_3}{\partial X_3} \end{vmatrix}. \tag{A.4}$$

To interpret the Jacobian in a physical way, consider a reference differential volume given by $dS^3 = d\omega$, where $d\boldsymbol{X}^{(1)} = dS\,\boldsymbol{e}_1$, $d\boldsymbol{X}^{(2)} = dS\,\boldsymbol{e}_2$, and $d\boldsymbol{X}^{(3)} = dS\,\boldsymbol{e}_3$. The current differential element is described by $d\boldsymbol{x}^{(1)} = \frac{\partial x_k}{\partial X_1} dS\,\boldsymbol{e}_k$, $d\boldsymbol{x}^{(2)} = \frac{\partial x_k}{\partial X_2} dS\,\boldsymbol{e}_k$, and $d\boldsymbol{x}^{(3)} = \frac{\partial x_k}{\partial X_3} dS\,\boldsymbol{e}_k$, where $\boldsymbol{e}$ is a unit vector, and

$$\underbrace{d\boldsymbol{x}^{(1)} \cdot (d\boldsymbol{x}^{(2)} \times d\boldsymbol{x}^{(3)})}_{\stackrel{\text{def}}{=} d\omega} = \begin{vmatrix} dx_1^{(1)} & dx_2^{(1)} & dx_3^{(1)} \\ dx_1^{(2)} & dx_2^{(2)} & dx_3^{(2)} \\ dx_1^{(3)} & dx_2^{(3)} & dx_3^{(3)} \end{vmatrix} = \begin{vmatrix} \frac{\partial x_1}{\partial X_1} & \frac{\partial x_2}{\partial X_1} & \frac{\partial x_3}{\partial X_1} \\ \frac{\partial x_1}{\partial X_2} & \frac{\partial x_2}{\partial X_2} & \frac{\partial x_3}{\partial X_2} \\ \frac{\partial x_1}{\partial X_3} & \frac{\partial x_2}{\partial X_3} & \frac{\partial x_3}{\partial X_3} \end{vmatrix} dS^3. \tag{A.5}$$

Therefore, $d\omega = J\, d\omega_0$. Thus, the Jacobian of the deformation gradient must remain positive definite; otherwise we obtain physically impossible "negative" volumes.

A.3 Equilibrium/kinetics of solid continua

We start with the following postulated balance law for an arbitrary part ω around a point P with boundary $\partial\omega$ of a body Ω:

$$\underbrace{\int_{\partial\omega} \boldsymbol{t}\, da}_{\text{surface forces}} + \underbrace{\int_{\omega} \boldsymbol{f}\, d\omega}_{\text{body forces}} = \underbrace{\frac{d}{dt}\int_{\omega} \rho \dot{\boldsymbol{u}}\, d\omega}_{\text{inertial forces}}, \tag{A.6}$$

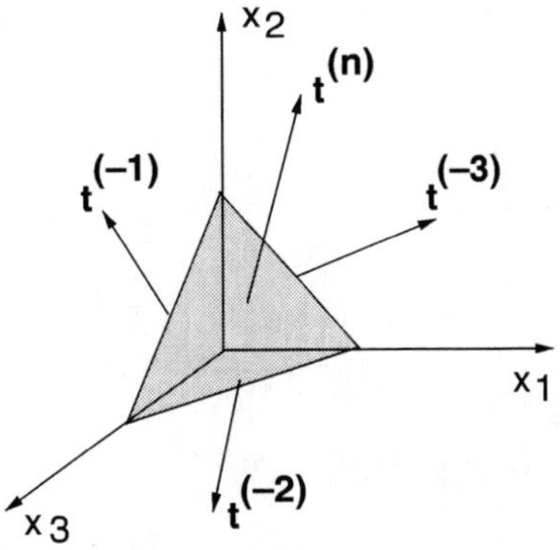

Figure A.1. *Cauchy tetrahedron: A "sectioned material point."*

where ρ is the material density, $\boldsymbol{b}$ is the body force per unit mass ($\boldsymbol{f} = \rho\boldsymbol{b}$), and $\dot{\boldsymbol{u}}$ is the time derivative of the displacement.[71]

When the actual molecular structure is considered on a submicroscopic scale, the force densities, $\boldsymbol{t}$, which we commonly refer to as "surface forces," are taken to involve short-range intermolecular forces. Tacitly we assume that the effects of radiative forces, and others that do not require momentum transfer through a continuum, are negligible. This is a so-called local action postulate. As long as the volume element is large, our resultant body and surface forces may be interpreted as sums of these intermolecular forces. When we pass to larger scales, we can justifiably use the continuum concept.

A.4 Postulates on volume and surface quantities

Consider a tetrahedron in equilibrium, as shown in Figure A.1. From Newton's laws,

$$\boldsymbol{t}^{(n)}\Delta A^{(n)} + \boldsymbol{t}^{(-1)}\Delta A^{(1)} + \boldsymbol{t}^{(-2)}\Delta A^{(2)} + \boldsymbol{t}^{(-3)}\Delta A^{(3)} + \boldsymbol{f}\Delta\Omega = \rho\Delta\Omega\ddot{\boldsymbol{u}},$$

where $\Delta A^{(n)}$ is the surface area of the face of the tetrahedron with normal $\boldsymbol{n}$ and $\Delta\Omega$ is the tetrahedron volume. Clearly, as the distance between the tetrahedron base (located at (0, 0, 0)) and the surface center, denoted by h, goes to zero, we have $h \to 0 \Rightarrow \Delta A^{(n)} \to 0 \Rightarrow \frac{\Delta\Omega}{\Delta A^{(n)}} \to 0$. Geometrically, we have $\frac{\Delta A^{(i)}}{\Delta A^{(n)}} = \cos(x_i, x_n) \stackrel{\text{def}}{=} n_i$, and therefore $\boldsymbol{t}^{(n)} + \boldsymbol{t}^{(-1)}\cos(x_1, x_n) + \boldsymbol{t}^{(-2)}\cos(x_2, x_n) + \boldsymbol{t}^{(-3)}\cos(x_3, x_n) = \boldsymbol{0}$.

It is clear that forces on the surface areas can be decomposed into three linearly independent components. It is convenient to pictorially represent the concept of stress at a point, representing the surface forces there, by a cube surrounding a point. The fundamental issue that must be resolved is the characterization of these surface forces. We can represent the force density vector, the so-called traction, on a surface by the component representation $\boldsymbol{t}^{(i)} \stackrel{\text{def}}{=} (\sigma_{1i}, \sigma_{2i}, \sigma_{3i})^T$, where the second index represents the direction of the component and the first index represents the normal to the corresponding coordinate plane. From this point forth, we will drop the superscript notation of $\boldsymbol{t}^{(n)}$, where it is implicit that $\boldsymbol{t} \stackrel{\text{def}}{=} \boldsymbol{t}^{(n)} = \boldsymbol{\sigma}^T \cdot \boldsymbol{n}$

[71] We use the shorthand notation $\dot{()} \stackrel{\text{def}}{=} \frac{d()}{dt}$.

or, explicitly ($t^{(1)} = -t^{(-1)}, t^{(2)} = -t^{(-2)}, t^{(3)} = -t^{(-3)}$),

$$t^{(n)} = t^{(1)}n_1 + t^{(2)}n_2 + t^{(3)}n_3 = \sigma^T \cdot n = \begin{bmatrix} \sigma_{11} & \sigma_{12} & \sigma_{13} \\ \sigma_{21} & \sigma_{22} & \sigma_{23} \\ \sigma_{31} & \sigma_{32} & \sigma_{33} \end{bmatrix}^T \begin{bmatrix} n_1 \\ n_2 \\ n_3 \end{bmatrix}, \tag{A.7}$$

where σ is the so-called Cauchy stress tensor.[72]

A.5 Balance law formulations

Substitution of Equation (A.5) into Equation (A.4) yields ($\omega \subset \Omega$)

$$\underbrace{\int_{\partial\omega} \sigma \cdot n \, da}_{\text{surface forces}} + \underbrace{\int_{\omega} f \, d\omega}_{\text{body forces}} = \underbrace{\frac{d}{dt}\int_{\omega} \rho \dot{u} \, d\omega}_{\text{inertial forces}}. \tag{A.8}$$

A relationship can be determined between the densities in the current and reference configurations: $\int_{\omega} \rho d\omega = \int_{\omega_0} \rho J d\omega_0 = \int_{\omega_0} \rho_0 d\omega_0$. Therefore, the Jacobian can also be interpreted as the ratio of material densities at a point. Since the volume is arbitrary, we can assume that $\rho J = \rho_0$ holds at every point in the body. Therefore, we may write $\frac{d}{dt}(\rho_0) = \frac{d}{dt}(\rho J) = 0$ when the system is mass conservative over time. This leads to writing the last term in Equation (A.6) as $\frac{d}{dt}\int_{\omega} \rho\dot{u} \, d\omega = \int_{\omega_0} \frac{d(\rho J)}{dt}\dot{u} \, d\omega_0 + \int_{\omega_0} \rho\ddot{u} J \, d\omega_0 = \int_{\omega} \rho\ddot{u} \, d\omega$. From Gauss's divergence theorem, and an implicit assumption that σ is differentiable, we have $\int_{\omega} (\nabla_x \cdot \sigma + f - \rho\ddot{u}) \, d\omega = \mathbf{0}$. If the volume is argued as being arbitrary, then the relation in the integral must hold pointwise, yielding

$$\nabla_x \cdot \sigma + f = \rho\ddot{u} = \rho\dot{v}, \tag{A.9}$$

where v is the velocity.

A.6 Symmetry of the stress tensor

Starting with an angular momentum balance, under the assumptions that no infinitesimal "micromoments" or so-called couple stresses exist, it can be shown that the stress tensor must be symmetric, i.e., $\int_{\partial\omega} x \times t \, da + \int_{\omega} x \times f \, d\omega = \frac{d}{dt}\int_{\omega} x \times \rho\dot{u} \, d\omega$, which implies $\sigma^T = \sigma$. It is somewhat easier to consider a differential element and to simply sum moments about the center. Doing this, one immediately obtains $\sigma_{12} = \sigma_{21}$, $\sigma_{23} = \sigma_{32}$, and $\sigma_{13} = \sigma_{31}$. Therefore,

$$t^{(n)} = t^{(1)}n_1 + t^{(2)}n_2 + t^{(3)}n_3 = \sigma \cdot n = \sigma^T \cdot n. \tag{A.10}$$

[72]Some authors follow the notation that the first index represents the direction of the component and the second index represents the normal to the corresponding coordinate plane. This leads to $t \stackrel{\text{def}}{=} t^{(n)} = \sigma \cdot n$. In the absence of couple stresses, a balance of angular momentum implies a symmetry of stress, $\sigma = \sigma^T$, and thus the difference in notations becomes immaterial.

A.7 The first law of thermodynamics

The interconversions of mechanical, thermal, and chemical energy in a system are governed by the first law of thermodynamics. It states that the time rate of change of the total energy, $\mathcal{K}+\mathcal{I}$, is equal to the sum of the work rate, $\mathcal{P}$, and the net heat supplied, $\mathcal{H}+\mathcal{Q}$:

$$\frac{d}{dt}(\mathcal{K}+\mathcal{I}) = \mathcal{P}+\mathcal{H}+\mathcal{Q}. \tag{A.11}$$

Here, the kinetic energy of a subvolume of material contained in Ω, denoted by ω, is $\mathcal{K} \stackrel{\text{def}}{=} \int_\omega \frac{1}{2}\rho\dot{\boldsymbol{u}}\cdot\dot{\boldsymbol{u}}\,d\omega$, the rate of work or power of external forces acting on ω is given by $\mathcal{P} \stackrel{\text{def}}{=} \int_\omega \rho\boldsymbol{b}\cdot\dot{\boldsymbol{u}}\,d\omega + \int_{\partial\omega}\boldsymbol{\sigma}\cdot\boldsymbol{n}\cdot\dot{\boldsymbol{u}}\,da$, the heat flow into the volume by conduction is $\mathcal{Q} \stackrel{\text{def}}{=} -\int_{\partial\omega}\boldsymbol{q}\cdot\boldsymbol{n}\,da = -\int_\omega \nabla_x\cdot\boldsymbol{q}\,d\omega$, the heat generated due to sources *such as chemical reactions* is $\mathcal{H} \stackrel{\text{def}}{=} \int_\omega \rho z\,d\omega$, and the stored energy is $\mathcal{I} \stackrel{\text{def}}{=} \int_\omega \rho w\,d\omega$. If we make the assumption that the mass in the system is constant, we have

$$\text{current mass} = \int_\omega \rho\,d\omega = \int_{\omega_0} \rho J\,d\omega_0 \approx \int_{\omega_0} \rho_0\,d\omega_0 = \text{original mass}, \tag{A.12}$$

which implies $\rho J = \rho_0$. Therefore, $\rho J = \rho_0 \Rightarrow \dot{\rho}J + \rho\dot{J} = 0$. Using this and the energy balance leads to

$$\begin{aligned}\frac{d}{dt}\int_\omega \frac{1}{2}\rho\dot{\boldsymbol{u}}\cdot\dot{\boldsymbol{u}}\,d\omega &= \int_{\omega_0}\frac{d}{dt}\frac{1}{2}(\rho J\dot{\boldsymbol{u}}\cdot\dot{\boldsymbol{u}})\,d\omega_0\\ &= \int_{\omega_0}\left(\frac{d}{dt}\rho_0\right)\frac{1}{2}\dot{\boldsymbol{u}}\cdot\dot{\boldsymbol{u}}\,d\omega_0 + \int_\omega \rho\frac{d}{dt}\frac{1}{2}(\dot{\boldsymbol{u}}\cdot\dot{\boldsymbol{u}})\,d\omega\\ &= \int_\omega \rho\dot{\boldsymbol{u}}\cdot\ddot{\boldsymbol{u}}\,d\omega.\end{aligned} \tag{A.13}$$

We also have

$$\frac{d}{dt}\int_\omega \rho w\,d\omega = \frac{d}{dt}\int_{\omega_0}\rho J w\,d\omega_0 = \int_{\omega_0}\frac{d}{dt}(\rho_0)w\,d\omega_0 + \int_\omega \rho\dot{w}\,d\omega. \tag{A.14}$$

By using the divergence theorem, we obtain

$$\int_{\partial\omega}\boldsymbol{\sigma}\cdot\boldsymbol{n}\cdot\dot{\boldsymbol{u}}\,da = \int_\omega \nabla_x\cdot(\boldsymbol{\sigma}\cdot\dot{\boldsymbol{u}})\,d\omega = \int_\omega(\nabla_x\cdot\boldsymbol{\sigma})\cdot\dot{\boldsymbol{u}}\,d\omega + \int_\omega \boldsymbol{\sigma}:\nabla_x\dot{\boldsymbol{u}}\,d\omega. \tag{A.15}$$

Combining the results, and enforcing balance of momentum, leads to

$$\begin{aligned}&\int_\omega(\rho\dot{w} + \dot{\boldsymbol{u}}\cdot(\rho\ddot{\boldsymbol{u}} - \nabla_x\cdot\boldsymbol{\sigma} - \rho\boldsymbol{b}) - \boldsymbol{\sigma}:\nabla_x\dot{\boldsymbol{u}} + \nabla_x\cdot\boldsymbol{q} - \rho z)\,d\omega\\ &\quad= \int_\omega(\rho\dot{w} - \boldsymbol{\sigma}:\nabla_x\dot{\boldsymbol{u}} + \nabla_x\cdot\boldsymbol{q} - \rho z)\,d\omega = 0.\end{aligned} \tag{A.16}$$

Since the volume ω is arbitrary, the integrand must hold locally and we have

$$\rho\dot{w} - \boldsymbol{\sigma}:\nabla_x\dot{\boldsymbol{u}} + \nabla_x\cdot\boldsymbol{q} - \rho z = 0. \tag{A.17}$$

A.8 Basic constitutive assumptions for fluid mechanics

A fluid at rest cannot support shear loading. This is the primary difference between a fluid and a solid. Therefore, for a fluid at rest, one can write

$$\boldsymbol{\sigma} = -P_o \mathbf{1}, \tag{A.18}$$

where $P_o = -\frac{\mathrm{tr}\,\boldsymbol{\sigma}}{3}$ is the hydrostatic pressure. In other words, there are no shear stresses in a fluid at rest.

In the dynamic case, the pressure, called the thermodynamic pressure, is related to the temperature and the fluid density by an equation of state

$$\mathcal{Z}(P, \rho, \theta) = 0. \tag{A.19}$$

For a fluid in motion,

$$\boldsymbol{\sigma} = -P\mathbf{1} + \boldsymbol{\tau}, \tag{A.20}$$

where $\boldsymbol{\tau}$ is a so-called viscous stress tensor.[73] Thus, for a compressible fluid in motion,

$$\frac{\mathrm{tr}\,\boldsymbol{\sigma}}{3} = -P + \frac{\mathrm{tr}\,\boldsymbol{\tau}}{3}. \tag{A.21}$$

In general, for a fluid we have

$$\boldsymbol{\tau} = \mathcal{G}(\boldsymbol{D}) \quad \text{and} \quad \boldsymbol{D} \stackrel{\text{def}}{=} \frac{1}{2}(\nabla_{\boldsymbol{x}}\boldsymbol{v} + (\nabla_{\boldsymbol{x}}\boldsymbol{v})^T), \tag{A.22}$$

where $\boldsymbol{v} = \dot{\boldsymbol{u}}$ is the velocity and $\boldsymbol{D}$ is the symmetric part of the velocity gradient. A Newtonian fluid is one where a linear relation exists between the viscous stresses and $\boldsymbol{D}$:

$$\boldsymbol{\tau} = \boldsymbol{V} : \boldsymbol{D}, \tag{A.23}$$

where $\boldsymbol{V}$ is a symmetric positive-definite (fourth-order) viscosity tensor. For an isotropic (standard) Newtonian fluid, we have

$$\boldsymbol{\sigma} = -P\mathbf{1} + \lambda_v \mathrm{tr}\,\boldsymbol{D}\mathbf{1} + 2\mu_v\boldsymbol{D} = -P\mathbf{1} + 3\kappa_v\frac{\mathrm{tr}\,\boldsymbol{D}}{3}\mathbf{1} + 2\mu_v\boldsymbol{D}', \tag{A.24}$$

where κ_v is called the bulk viscosity, λ_v is a viscosity constant, and μ_v is the shear viscosity. Explicitly, with an (x, y, z) Cartesian triad,

$$\underbrace{\begin{Bmatrix} \sigma_{xx} \\ \sigma_{yy} \\ \sigma_{zz} \\ \sigma_{xy} \\ \sigma_{yz} \\ \sigma_{zx} \end{Bmatrix}}_{\stackrel{\text{def}}{=}\{\boldsymbol{\sigma}\}} = \underbrace{\begin{Bmatrix} -P \\ -P \\ -P \\ 0 \\ 0 \\ 0 \end{Bmatrix}}_{\stackrel{\text{def}}{=}\{-\boldsymbol{P}\}} + \underbrace{\begin{bmatrix} c_1 & c_2 & c_2 & 0 & 0 & 0 \\ c_2 & c_1 & c_2 & 0 & 0 & 0 \\ c_2 & c_2 & c_1 & 0 & 0 & 0 \\ 0 & 0 & 0 & \mu_v & 0 & 0 \\ 0 & 0 & 0 & 0 & \mu_v & 0 \\ 0 & 0 & 0 & 0 & 0 & \mu_v \end{bmatrix}}_{\stackrel{\text{def}}{=}[\boldsymbol{V}]} \underbrace{\begin{Bmatrix} D_{xx} \\ D_{yy} \\ D_{zz} \\ 2D_{xy} \\ 2D_{yz} \\ 2D_{zx} \end{Bmatrix}}_{\stackrel{\text{def}}{=}\{\boldsymbol{D}\}}, \tag{A.25}$$

[73] An inviscid or "perfect" fluid is one where $\boldsymbol{\tau}$ is taken to be zero, even when motion is present.

where $c_1 = \kappa_v + \frac{4}{3}\mu_v$ and $c_2 = \kappa_v - \frac{2}{3}\mu_v$, $D_{xx} = \frac{\partial v_x}{\partial x}$, $D_{yy} = \frac{\partial v_y}{\partial y}$, $D_{zz} = \frac{\partial v_z}{\partial z}$, and

$$D_{xy} = \frac{1}{2}\left(\frac{\partial v_x}{\partial y} + \frac{\partial v_y}{\partial x}\right), \quad D_{yz} = \frac{1}{2}\left(\frac{\partial v_y}{\partial z} + \frac{\partial v_z}{\partial y}\right), \quad D_{zx} = \frac{1}{2}\left(\frac{\partial v_z}{\partial x} + \frac{\partial v_x}{\partial z}\right). \tag{A.26}$$

The so-called Stokes condition attempts to force the thermodynamic pressure to collapse to the classical definition of mechanical pressure, i.e.,

$$\frac{\operatorname{tr}\boldsymbol{\sigma}}{3} = -P + 3\kappa_v \frac{\operatorname{tr}\boldsymbol{D}}{3} = -P, \tag{A.27}$$

leading to the conclusion that $\kappa_v = 0$ or $\lambda_v = -\frac{2}{3}\mu_v$. Thus, a Newtonian fluid obeying the Stokes condition has the following constitutive law:

$$\boldsymbol{\sigma} = -P\mathbf{1} - \frac{2}{3}\mu_v \operatorname{tr}\boldsymbol{D}\mathbf{1} + 2\mu_v \boldsymbol{D} = -P\mathbf{1} + 2\mu_v \boldsymbol{D}'. \tag{A.28}$$

From the conservation of mass relation derived earlier, we have

$$\frac{d}{dt}(\rho_0) = \frac{d}{dt}(\rho J) = J\frac{d\rho}{dt} + \rho\frac{dJ}{dt} = 0, \tag{A.29}$$

which leads to

$$\frac{d\rho}{dt} + \frac{\rho}{J}\frac{dJ}{dt} = 0. \tag{A.30}$$

Since

$$\dot{J} = \frac{d}{dt}\det\boldsymbol{F} = (\det\boldsymbol{F})\operatorname{tr}(\dot{\boldsymbol{F}}\cdot\boldsymbol{F}^{-1}) = J\operatorname{tr}\boldsymbol{L}, \tag{A.31}$$

where $\boldsymbol{L} = \nabla_{\boldsymbol{x}}\boldsymbol{v}$ is the velocity gradient, Equation (A.29) becomes

$$\frac{d\rho}{dt} + \rho\nabla_{\boldsymbol{x}}\cdot\boldsymbol{v} = 0. \tag{A.32}$$

Now we write the total temporal ("material") derivative in convective form:

$$\frac{d\rho}{dt} = \frac{\partial\rho}{\partial t} + (\nabla_{\boldsymbol{x}}\rho)\cdot\frac{d\boldsymbol{x}}{dt} = \frac{\partial\rho}{\partial t} + \nabla_{\boldsymbol{x}}\rho\cdot\boldsymbol{v}. \tag{A.33}$$

Thus, Equation (A.32) becomes

$$\frac{\partial\rho}{\partial t} + \nabla_{\boldsymbol{x}}\rho\cdot\boldsymbol{v} + \rho\nabla_{\boldsymbol{x}}\cdot\boldsymbol{v} = \frac{\partial\rho}{\partial t} + \nabla_{\boldsymbol{x}}\cdot(\rho\boldsymbol{v}) = 0. \tag{A.34}$$

Thus, in summary, the coupled governing equations are

$$\boxed{\begin{aligned} &\mathcal{Z}(P,\rho,\theta) = 0, \\ &\frac{\partial\rho}{\partial t} = -\nabla_{\boldsymbol{x}}\cdot(\rho\boldsymbol{v}), \\ &\rho\dot{w} = \boldsymbol{\sigma} : \nabla_{\boldsymbol{x}}\boldsymbol{v} - \nabla_{\boldsymbol{x}}\cdot\boldsymbol{q} + \rho z, \\ &\rho\dot{\boldsymbol{v}} = \nabla_{\boldsymbol{x}}\cdot\boldsymbol{\sigma} + \rho\boldsymbol{b}. \end{aligned}} \tag{A.35}$$

Collectively, we refer to these equations as the Navier–Stokes equations.

Remark. It is usually helpful to write both of the total time derivatives appearing above as

$$\boxed{\begin{aligned} \frac{d\boldsymbol{v}}{dt} &= \left.\frac{\partial \boldsymbol{v}}{\partial t}\right|_{\boldsymbol{x}} + (\nabla_{\boldsymbol{x}}\boldsymbol{v})\big|_t \cdot \frac{d\boldsymbol{x}}{dt}, \\ \frac{d\theta}{dt} &= \left.\frac{\partial \theta}{\partial t}\right|_{\boldsymbol{x}} + (\nabla_{\boldsymbol{x}}\theta)\big|_t \cdot \frac{d\boldsymbol{x}}{dt}, \end{aligned}} \tag{A.36}$$

thus leading to (with $w = C\theta$ and $\boldsymbol{q} = -\mathbb{K}\cdot\nabla_{\boldsymbol{x}}\theta$)

$$\boxed{\begin{aligned} &\frac{\partial \rho}{\partial t} = -\nabla_{\boldsymbol{x}}\rho\cdot\boldsymbol{v} - \rho\nabla_{\boldsymbol{x}}\cdot\boldsymbol{v}, \\ &\rho C\left(\frac{\partial \theta}{\partial t} + (\nabla_{\boldsymbol{x}}\theta)\cdot\boldsymbol{v}\right) = \boldsymbol{\sigma} : \nabla_{\boldsymbol{x}}\boldsymbol{v} + \nabla_{\boldsymbol{x}}\cdot\mathbb{K}\cdot\nabla_{\boldsymbol{x}}\theta + \rho z, \\ &\rho\left(\frac{\partial \boldsymbol{v}}{\partial t} + (\nabla_{\boldsymbol{x}}\boldsymbol{v})\cdot\boldsymbol{v}\right) = \nabla_{\boldsymbol{x}}\cdot\boldsymbol{\sigma} + \rho\boldsymbol{b}, \\ &\boldsymbol{\sigma} = -P\mathbf{1} + \lambda_v \mathrm{tr}\,\boldsymbol{D}\mathbf{1} + 2\mu_v\boldsymbol{D} = -P\mathbf{1} + 3\kappa_v\frac{\mathrm{tr}\boldsymbol{D}}{3}\mathbf{1} + 2\mu_v\boldsymbol{D}', \end{aligned}} \tag{A.37}$$

where, for example, P is given by Equation (8.49).

Remark. When the Navier–Stokes equations are put into nondimensional form, several nondimensional numbers appear. Most prominent is the Reynolds number, which measures the inertial forces relative to the viscous forces:

$$Re \stackrel{\text{def}}{=} \frac{\rho v L}{\mu}, \tag{A.38}$$

where L is an intrinsic length scale in the system. High Reynolds numbers usually lead to turbulent flows where the Newtonian fluid hypothesis is questionable. Constitutive laws that are applicable in a truly turbulent regime are beyond the scope of this work.

Appendix B
Scattering

B.1 Generalized Fresnel relations

In order to further illustrate the dependency of the results on $\hat{n}$, recall the fundamental relation for reflectance

$$R = \frac{1}{2}\left(\left(\frac{\frac{\hat{n}^2}{\hat{\mu}}\cos\theta_i - (\hat{n}^2 - \sin^2\theta_i)^{\frac{1}{2}}}{\frac{\hat{n}^2}{\hat{\mu}}\cos\theta_i + (\hat{n}^2 - \sin^2\theta_i)^{\frac{1}{2}}}\right)^2 + \left(\frac{\cos\theta_i - \frac{1}{\hat{\mu}}(\hat{n}^2 - \sin^2\theta_i)^{\frac{1}{2}}}{\cos\theta_i + \frac{1}{\hat{\mu}}(\hat{n}^2 - \sin^2\theta_i)^{\frac{1}{2}}}\right)^2\right), \tag{B.1}$$

whose variation as a function of the angle θ_i is depicted in Figure B.1. For all but $\hat{n} = 2$, there is discernible nonmonotone behavior. The nonmonotone behavior is slight for $\hat{n} = 4$, but nonetheless present. Clearly, as $\hat{n} \to \infty$, $R \to 1$, no matter what the angle of incidence's value. Also, as $\hat{n} \to 1$, provided that $\hat{\mu} = 1$, $R \to 0$, i.e., all incident energy is absorbed. With increasing $\hat{n}$, the angle for minimum reflectance grows larger. Figure B.1 illustrates the behavior for $\hat{\mu} = 1$. For $\hat{\mu} \neq 1$, see Figure B.2, which illustrates the variation of R when $\hat{\mu} = 2$ and $\hat{\mu} = 10$.

B.2 Biological applications: Multiple red blood cell light scattering

Erythrocytes or red blood cells (RBCs) are the most numerous cells in human blood and are responsible for the transport of oxygen and carbon dioxide. Typically, at a standard altitude, healthy females average about 4.8 million of these cells per cubic millimeter of blood, while healthy males average about 5.4 million per cubic millimeter. The lifespan of RBCs is approximately 120 days. Thereafter, they are ingested by phagocytic cells in the liver and spleen (approximately 3 million RBCs die and are scavenged each second), and the iron in their hemoglobin (which gives them their characteristic dark color) is reclaimed for reuse. The remainder of the heme portion of the molecule is degraded into bile pigments and excreted by the liver. The typical biconcaval shape of RBCs is the optimal combination of surface area to volume ratio. This shape also provides unique deformability characteristics

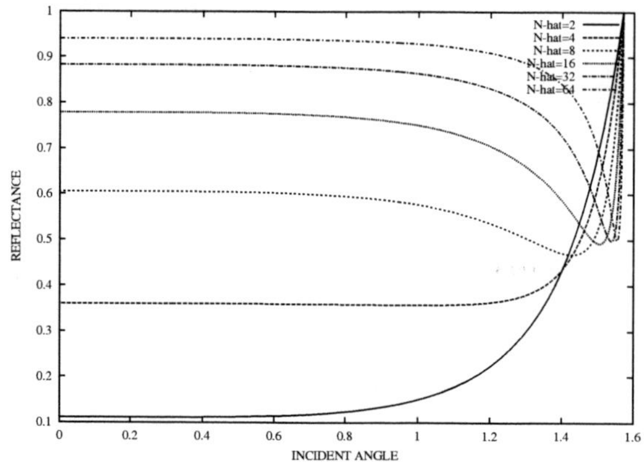

Figure B.1. *The variation of the reflectance, R, with angle of incidence. For all but $\hat{n} = 2$, there is discernible nonmonotone behavior. The behavior is slight for $\hat{n} = 4$, but nonetheless present (Zohdi* [219]*).*

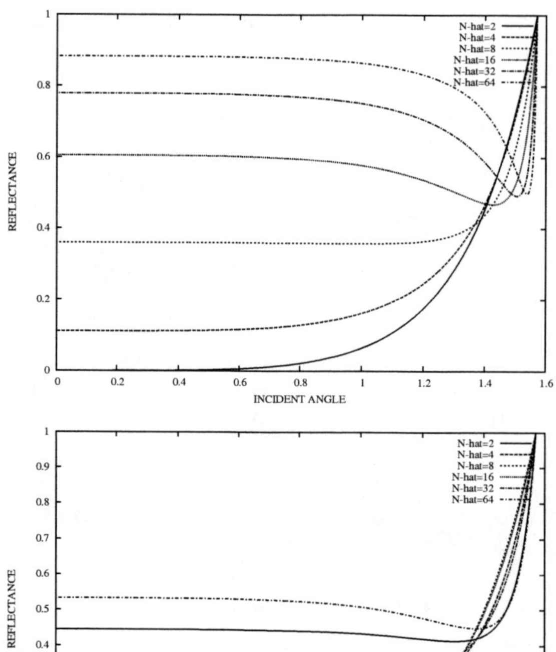

Figure B.2. *The variation of the reflectance, R, with angle of incidence for $\hat{\mu} = 2$ (top) and $\hat{\mu} = 10$ (bottom) (Zohdi* [219]*).*

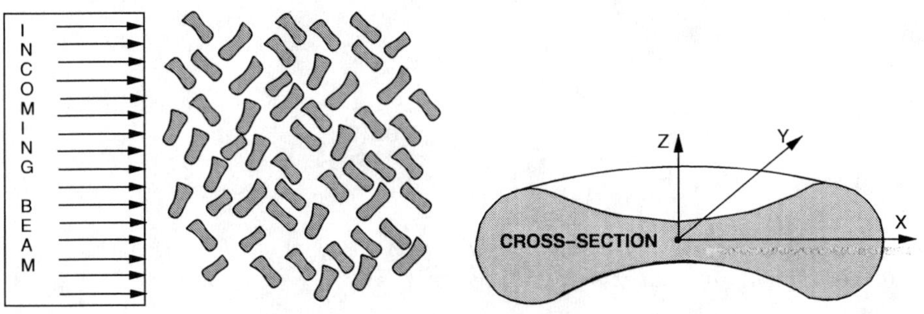

Figure B.3. *Left, the scattering system considered, comprising a beam, made up of multiple rays, incident on a collection of randomly distributed RBCs. Right, a typical RBC (Zohdi and Kuypers* [223]*).*

to the cell, giving it advantageous properties in order to perform its function in small capillaries. Deviation from the usual healthy cell morphology can lead to a loss of normal function and reduced RBC survival. Hence, measurement of RBC shape is an important parameter for describing RBC function.

A significant part of determining the characteristics of blood is achieved via optical measurements. Ideally, one would like to perform numerical simulations in order to minimize time-consuming laboratory tests. Accordingly, the objective of this work is to develop a simple approach to ascertaining the light-scattering response of large numbers of randomly distributed and oriented RBCs. Because the diameter of a typical RBC is on the order of eight microns ($d \approx 8 \times 10^{-6}$ m), which is much larger than the wavelengths of visible light (approximately $3.8 \times 10^{-7}\ \mathrm{m} \leq \lambda \leq 7.8 \times 10^{-7}$ m), geometric ray-tracing can be used to determine the amount of propagating optical energy, characterized by the Poynting vector, that is reflected and absorbed by multiple RBCs.[74] Ray-tracing is highly amenable to the rapid large-scale computation needed to track the scattering of incident light beams, comprising multiple rays, by multiple cells (Figure B.3), thus making it an ideal simulation paradigm.

The specific model problem that we consider is an initially coherent beam (Figure B.3), composed of multiple collinear rays, where each ray is a vector in the direction of the flow of electromagnetic (optical) energy, which, in isotropic media, corresponds to the normal to the wave front. Thus, for isotropic media, the rays are parallel to the wave's propagation vector (Figure B.3). Of particular interest is to describe the breakup of initially highly directional coherent beams, for example, lasers, which do not spread out into multidirectional rays unless they encounter multiple scatterers. The overall objective of this section is to provide a straightforward approach that can be implemented by researchers in the field, using standard desktop computers.

[74] See Hecht [91], Born and Wolf [35], Gross [86], Bohren and Huffman [33], Elmore and Heald [63], and van de Hulst [197].

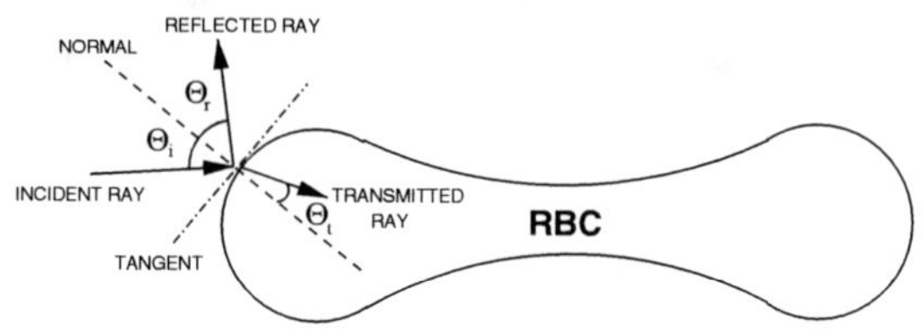

Figure B.4. *The nomenclature for Fresnel's equations for an incident ray that encounters a scattering cell (Zohdi and Kuypers* [223]*).*

B.2.1 Parametrization of cell configurations

One of the most widely cited biconcaval representations for RBCs (Figure B.3) is (Evans and Fung [64])

$$F \stackrel{\text{def}}{=} \left(\frac{2(z-z_o)}{b}\right)^2 - \left(1 - \frac{(x-x_o)^2+(y-y_o)^2}{b^2}\right) \times \left(c_o + c_1\left(\frac{(x-x_o)^2+(y-y_o)^2}{b^2}\right) + c_2\left(\frac{(x-x_o)^2+(y-y_o)^2}{b^2}\right)^2\right)^2 = 0. \tag{B.2}$$

The outward surface normals, $\boldsymbol{n}$, needed later during the scattering calculations (Figure B.4), are easy to characterize by computing $\boldsymbol{n} = \frac{\nabla F}{||\nabla F||}$. The orientation of the cells, usually random, can be controlled, via standard rotational coordinate transformations, with random angles (Figure B.4).

The classical random sequential addition algorithm (Widom [200]) is used to place nonoverlapping cells randomly into the domain of interest. This algorithm is adequate for the volume fraction range of interest. However, if higher volume fractions are desired, more sophisticated algorithms, such as the equilibrium-based Metropolis algorithm, can be used. See Torquato [194] for a detailed review of such methods. Furthermore, for much higher volume fractions, effectively packing (and "jamming") particles to theoretical limits, a new class of methods, based on simultaneous particle flow and growth, has been developed by Torquato and coworkers (see, for example, Kansaal et al. [119] and Donev et al. [55]–[59]).

Remark. Henceforth, we assume that the medium surrounding the cells behaves as a vacuum; thus, there are no energetic losses as the electromagnetic rays pass through it. Furthermore, we assume that all electromagnetic energy that is absorbed by a cell becomes trapped and is not re-emitted. This assumption is discussed further later.

B.2.2 Computational algorithm

The primary quantity of interest is the behavior of the propagation of the optical energy, characterized by the irradiance. For example, consider the following metrics for overall irradiance of the beam:

$$I_x \stackrel{\text{def}}{=} \frac{1}{I_o}\sum_{i=1}^{N_r} \boldsymbol{S}_i \cdot \boldsymbol{e}_x, \quad I_y \stackrel{\text{def}}{=} \frac{1}{I_o}\sum_{i=1}^{N_r} \boldsymbol{S}_i \cdot \boldsymbol{e}_y, \quad \text{and} \quad I_z \stackrel{\text{def}}{=} \frac{1}{I_o}\sum_{i=1}^{N_r} \boldsymbol{S}_i \cdot \boldsymbol{e}_z, \tag{B.3}$$

(1) COMPUTE RAY REFLECTIONS (FRESNEL RELATIONS);
(2) COMPUTE ABSORPTION BY CELLS;
(3) INCREMENT ALL RAY POSITIONS: $\boldsymbol{r}_i(t+\Delta t) = \boldsymbol{r}_i(t) + \Delta t \boldsymbol{v}_i(t), \quad i = 1, \ldots, RAYS$;
(4) GO TO (1) AND REPEAT WITH $t = t + \Delta t$.

ALGORITHM B.1

where N_r is the number of rays making up the beam and $I_o = ||\boldsymbol{I}(0)||$ is the magnitude of the initial irradiance at time $t = 0$. The computational algorithm is given as Algorithm B.1, starting at $t = 0$ and ending at $t = T$.

Remark. The time step size Δt is dictated by the size of the cells. A somewhat ad hoc approach is to scale the time step size according to $\Delta t \propto \frac{\xi b}{||\boldsymbol{v}||}$, where b is the radius of the cells, $||\boldsymbol{v}||$ is the magnitude of the velocity of the rays, and ξ is a scaling factor; typically, $0.05 \leq \xi \leq 0.1$.

Remark. For step (1), it is convenient to determine whether a ray has just entered a cell domain by checking if $F(\hat{x}, \hat{y}, \hat{z}) \leq 0$, where $(\hat{x}, \hat{y}, \hat{z})$ are the coordinates of the cell expressed in a rotated frame that is aligned with the axes of symmetry of the cell, and then to compute the normal $\boldsymbol{n} = \frac{\nabla F}{||\nabla F||}$ in that frame.

B.2.3 A computational example

System parameters

We considered groups of randomly dispersed equal-sized cells, of increasing number, $N_c =$ 1000, 2000, 4000, and 8000, in a rectangular domain of dimensions (Figure B.5) 1 mm $\times$ 1 mm $\times$ 1 cm. This corresponds to a section of a standard testing device, described in detail in the next section. The stated number of cells corresponded to standard testing hematocrit values. The cells' major diameter was the nominal value of $d = 8 \times 10^{-6}$ m. A commonly used set of geometric parameters for the cell in Equation (B.2) is given by Evans and Fung [64] as $c_o = 0.207161$, $c_1 = 2.002558$, and $c_2 = -1.122762$. The beam was of circular cross section with diameter 0.79375 mm (1/32 of an inch, which falls in the range of beams used in experiments described later). The irradiance (Poynting vector magnitude) beam parameter was set to $I = I_o$ N $\cdot$ m/(m$^2\cdot$ s), where the irradiance for each ray was calculated as $I_o a_b / N_r$, where a_b was the cross-sectional area of the beam.[75] We used successively higher ray densities of $N_r = 200, 400, 600, 800, 1000$, etc., rays (Figure B.5) to represent the beam. The simulations were run until the rays completely exited the domain, which corresponded to a time scale on the order of $\frac{10^{-2}\,\text{m}}{c}$, where c is the speed of light. The initial velocity vector for all of the initially collinear rays making up the beam was $\boldsymbol{v} = (c, 0, 0)$.

[75] Because of the normalized structure of the metric, it is insensitive to the magnitude of I_o for the scattering calculations. The initial magnitude of the Poynting vector is $||\boldsymbol{I}(0)|| = \sqrt{I_x(0)^2 + I_y(0)^2 + I_z(0)^2}$, where, initially, only one component is nonzero, $I_x(0) = I_o$, in the x direction.

Figure B.5. *Starting from left to right and top to bottom, the progressive movement of rays* (1000) *making up a beam* ($\hat{n}$ = 1.075). *The lengths of the vectors indicate the irradiance (Zohdi and Kuypers* [223]). *The diameter* (8000 *cells) of the scatterers is given by Equation* (B.2).

Computational results

The ratio of the refractive indices $\hat{n}$ was chosen to vary around 1.0. The exact value corresponds to the state of the cell, including membrane characteristics and hemoglobin concentration. We chose a ratio of refractive indices of $\hat{n} \approx \frac{1.4}{1.3} \approx 1.075$, which is consistent with values commonly found in the literature. As the plots in Figure B.6 indicate, the total amount of energy that is forwardly scattered (defined as the component's Poynting ray vectors in the positive x direction) for $\hat{n} = 1.075$ decreases with the number of cells (scatterers).[76] A sequence of frames of the typical ray motion is provided in Figure B.5. Table B.1 tabulates the transmitted energy for various numbers of cells present. *It is important to emphasize that these calculations were performed within a few minutes on a single standard (DELL Precision* 3.3 GHz) *laptop.*

[76]The system at time $t = T$ indicated that all rays had exited the scattering system.

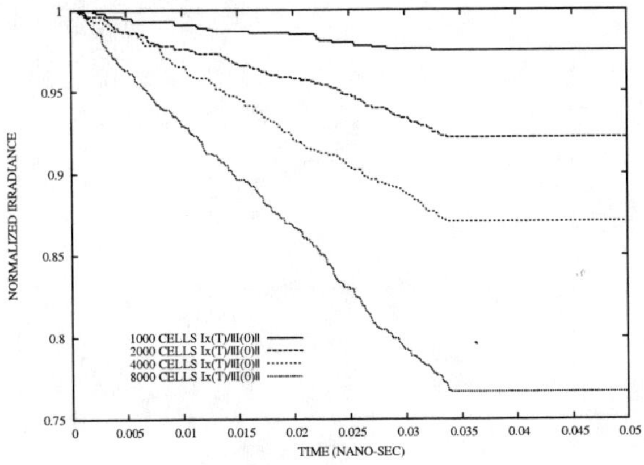

Figure B.6. *Computational results for the propagation of the forward scatter of $I_x(t)/||I(0)||$ for increasingly larger numbers of cells in the sample (Zohdi and Kuypers* [223]*).*

Table B.1. *Computational results for the forward scatter of $I_x(T)/||I(0)||$ (Zohdi and Kuypers* [223]*).*

Cells	$\frac{I_x(T)}{\|\|I(0)\|\|}$
1000	0.97501
2000	0.92201
4000	0.87046
8000	0.76656

Remark. Computational tests with higher ray resolution were also performed. We increased the ray density up to 10000 rays (starting from 200 rays), but found negligible change with respect to the 1000-ray resolution simulation. Thus, beyond $N_r = 1000$ rays, the computational results changed negligibly and can be considered to have converged. This cell/ray system provided stable results, i.e., increasing the number of rays and/or the number of cells surrounding the beam resulted in negligibly different overall system responses. Of course, there can be cases where much higher resolution is absolutely necessary. Thus, it is important to note that a straightforward, natural, algorithmic parallelism is possible with this computational technique. This can be achieved in two possible ways: (1) by assigning each processor its share of the rays and checking which cells make contact with those rays, or (2) by assigning each processor its share of particles and checking which rays make contact with those cells.

Laboratory experiments

Preparation of human and murine erythrocytes (RBC): Blood samples from healthy donors were collected in EDTA anticoagulant, after informed consent, at the Children's Hospital Oakland Research Institute (CHORI). Whole blood was kept at 4° C and used within 24 hours. RBCs were isolated by centrifugation, washed three times in HEPES-

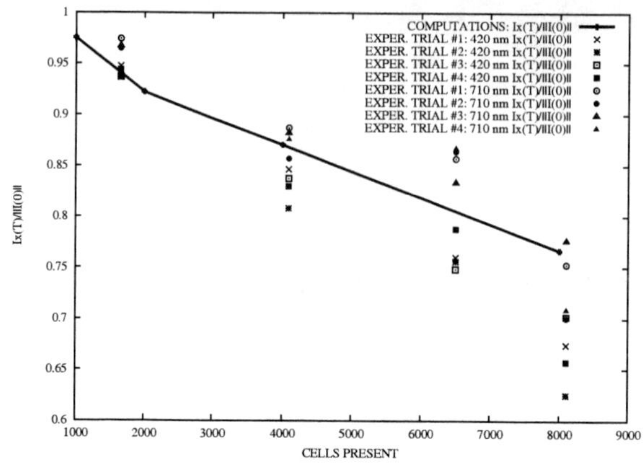

Figure B.7. *A comparison between the computational predictions and laboratory results for* 710-nm *and* 420-nm *light (four trials each, Zohdi and Kuypers* [223]*).*

buffered saline, and the buffy coat was removed after each wash. RBCs were resuspended at 30% hematocrit in HEPES buffered saline (150 mM NaCl, 10 mM HEPES, pH 7.4) and stored at 4° C until used within 48 hours. Before use, cells were suspended in buffer at room temperature to a cell concentration as indicated. The exact cell count in the suspension was determined using the Guava Easycount flowcytometer (Guava Technologies, Hayward, CA).

Light scatter measurements: 1.5 ml of cell suspension containing the indicated cell concentration in a cuvet with a 1-cm light path was put in a Varian 50 Cary Bio spectrophotometer (Varian Analytical Instruments, Palo Alto, CA). Light transmittance ($T = I_x/||I(0)||$), defined as the ratio of intensity of detected light (I_x) to incoming light ($||I(0)||$) of cell suspensions relative to buffer without cells, was recorded and averaged over a one minute interval. Wavelengths were varied from 200 to 800 nm as indicated and specific measurements were performed at 420 and 710 nm, the wavelengths of maximum and minimum light absorbance, respectively. In addition, the intensity of the incoming beam was restricted to approximately 1% of the original intensity by a neutral filter.

Comparison between computational predictions and experimental results

In the range of cell concentrations tested, the computational predictions and laboratory results are in close agreement, as indicated in Figure B.7 and Tables B.1, B.2, and B.3. Although the computations corresponded closely to both wavelengths of light, the match is closer to the 710-nm wavelength, since that wavelength reflects in a manner more consistent with the ratio of refractive indices used in the computations, as opposed to the 420-nm wavelength light, which is nearly a purely absorbing combination with RBCs.

Remark. Figure B.7 shows the relative light transmittance T as a function of the number of cells per milliliter for different wavelengths of light. Whereas the incoming light ($I(0)$) was greatly affected by placing masks with different circular cross sections in the light path, the transmittance T was not affected. The diameter of 1/32 of an inch for the beam

Table B.2. *Experimental results for the forward scatter of* $I_x(T)/||I(0)||$ *for* 420-nm *light (four trials).*

Cells	$\frac{I_x(T)}{\|\|I(0)\|\|}$: # 1	$\frac{I_x(T)}{\|\|I(0)\|\|}$: # 2	$\frac{I_x(T)}{\|\|I(0)\|\|}$: # 3	$\frac{I_x(T)}{\|\|I(0)\|\|}$: # 4
1650	0.94720	0.93630	0.93690	0.94360
4090	0.84640	0.80800	0.83740	0.82970
6510	0.75980	0.75610	0.74840	0.78770
8100	0.67440	0.62520	0.70220	0.65750

Table B.3. *Experimental results for the forward scatter of* $I_x(T)/||I(0)||$ *for* 710-nm *light (four trials).*

Cells	$\frac{I_x(T)}{\|\|I(0)\|\|}$: # 1	$\frac{I_x(T)}{\|\|I(0)\|\|}$: # 2	$\frac{I_x(T)}{\|\|I(0)\|\|}$: # 3	$\frac{I_x(T)}{\|\|I(0)\|\|}$: # 4
1650	0.97390	0.96450	0.96700	0.96760
4090	0.88700	0.85700	0.88230	0.87580
6510	0.85700	0.86390	0.83370	0.86710
8100	0.75300	0.70050	0.77650	0.70900

used for computation falls within the size used in our experimental approach. Furthermore, reducing the incoming light to 1% of its original value by the use of a neutral filter did not affect the transmittance. The data indicated in figures and tables were collected without restriction on the incoming light. Together, these data indicate that the beam intensity chosen for the computational model corresponded to the experimental approach.

Remark. We remark that, in the computations, the refracted energy absorbed by the cells was assumed to remain trapped within the cell. Certainly, some of the energy absorbed by the cells is converted into heat. An analysis of the thermal conversion process can be found in the main body of the monograph. Another level of complexity involves dispersion when light is transmitted through cells. Dispersion is the decomposition of light into its component wavelengths (or colors), which occurs because the index of refraction of a transparent medium is greater for light of shorter wavelengths. Accounting for dispersive effects is quite complex since it leads to a dramatic growth in the number of rays.

B.2.4 Extensions and concluding remarks

In summary, the objective of this section was to develop a simple computational framework, based on geometrical optics methods, to rapidly determine the light-scattering response of multiple RBCs. Because the wavelength of light (roughly 3.8×10^{-7} m $\leq \lambda \leq 7.8 \times 10^{-7}$ m) is approximately an order of magnitude smaller than the typical RBC scatterer ($d \approx 8 \times 10^{-6}$ m), geometric ray-tracing theory is applicable and can be used to rapidly ascertain the amount of propagating optical energy, characterized by the Poynting vector, that is reflected and absorbed by multiple cells. Three-dimensional examples were given to illustrate the technique, and the computational results match closely with experiments performed on blood samples at the red cell laboratory at CHORI.

We conclude by stressing a few points for possible extensions. First, a more general way to characterize a wider variety of RBC states, which are not necessarily always biconcaval, can be achieved by modifying the equation for a generalized "hyper"-ellipsoid:

$$F \stackrel{\text{def}}{=} \left(\frac{|x - x_o|}{r_1}\right)^{s_1} + \left(\frac{|y - y_o|}{r_2}\right)^{s_2} + \left(\frac{|z - z_o|}{r_3}\right)^{s_3} = 1, \tag{B.4}$$

where the s's are exponents. Values of $s < 1$ produce nonconvex shapes, while $s > 2$ values produce "block-like" shapes. Furthermore, we can introduce the particulate aspect ratio, defined by $AR \stackrel{\text{def}}{=} \frac{r_1}{r_2} = \frac{r_1}{r_3}$, where $r_2 = r_3$, $AR > 1$ for prolate geometries, and $AR < 1$ for oblate shapes. To produce the shape of a typical RBC, we introduce an extra term in the denominator of the first axis term:

$$F \stackrel{\text{def}}{=} \left(\frac{|x - x_o|}{r_1 + c_1\lambda^{c_2}}\right)^{s_1} + \left(\frac{|y - y_o|}{r_2}\right)^{s_2} + \left(\frac{|z - z_o|}{r_3}\right)^{s_3} = 1, \tag{B.5}$$

where $\lambda = \sqrt{y^2 + z^2}$ and $c_1 \geq 0$ and $c_2 \geq 0$. The effect of the term $c_1\lambda^{c_2}$ is to make the effective radius of the ellipsoid in the x direction grow as one moves away from the origin. As before, the outward surface normals $\boldsymbol{n}$ needed during the scattering calculations are easy to characterize by writing $\boldsymbol{n} = \frac{\nabla F}{||\nabla F||}$ with respect to a rotated frame that is aligned with the axes of symmetry of the generalized cell.

Second, it is important to recognize that one can describe the aggregate ray behavior in a more detailed manner via higher moment distributions of the individual ray fronts and their velocities. For example, consider any quantity Q with a distribution of values ($Q_i, i = 1, 2, \ldots, N_r$ = rays) about an arbitrary reference value, denoted by $Q^\star$, as $\boldsymbol{M}_p^{Q_i - Q^\star} \stackrel{\text{def}}{=} \frac{\sum_{i=1}^{N_r}(Q_i - Q^\star)^p}{N_r}$, where $A \stackrel{\text{def}}{=} \frac{\sum_{i=1}^{N_r} Q_i}{N_r}$. The various moments characterize the distribution. For example, (I) $\boldsymbol{M}_1^{Q_i - A}$ measures the first deviation from the average, which equals zero, (II) $\boldsymbol{M}_1^{Q_i - 0}$ is the average, (III) $\boldsymbol{M}_2^{Q_i - A}$ is the standard deviation, (IV) $\boldsymbol{M}_3^{Q_i - A}$ is the skewness, and (V) $\boldsymbol{M}_4^{Q_i - A}$ is the kurtosis. The higher moments, such as the skewness, measure the bias, or asymmetry, of the distribution of data, while the kurtosis measures the degree of peakedness of the distribution of data around the average.

Finally, when more microstructural features are considered, for example, topological and thermal variables, parameter studies become quite involved. In order to eliminate a trial and error approach to determining the characteristics of the types of cells that would be needed to achieve a certain level of scattering, the genetic algorithms presented earlier can be used to ascertain scatterer combinations that deliver prespecified electromagnetic scattering, thermal responses, and radiative (infrared) emission.

Generally, RBC behavior under fluid shear stress and response to osmolality changes is essential for normal function and survival. The ability to predict and measure the shape and deformation of individual RBCs under fluid shear stress will improve diagnosis of RBC disorders and open new avenues to treatment. New nanotechnology approaches coupled with real-time computational analysis will make it feasible to generate shape and deformability histograms in very small volumes of blood. This line of research is currently being pursued by the author, in particular to help detect blood disorders, which are characterized by the deviation of the shape of cells from those of healthy ones under standard test conditions. Such disorders, in theory, could be detected by differences in their scattering responses from those of healthy cells.

Red cell shape is essential for proper circulation. Changes in shape will lead to decreased red cell survival, often accompanied by anemia. Genetic disorders of cytoskeletal proteins will lead to red cell pathology, including hereditary spherocytosis and hereditary elliptocytosis (Eber and Lux [62] and Gallagher [73], [74]). Changes in membrane and cytosolic proteins may affect the state of hydration of the cell and thereby its morphology. Millions of humans are affected by hemoglobinopathies such as sickle-cell disease and thalassemia (Forget and Cohen [69] and Steinberg et al. [178]). The altered hemoglobin in these disorders can lead to changes in red cell properties, including membrane damage. Any of these conditions will result in an alteration of the scattering properties of the population of red cells. It is hoped that simple scatter measurements and fitting of the obtained data to our simulation model will reveal altered parameters of the red cell population related to red cell pathology. We hypothesize that this approach may be used as part of the diagnostic process or to evaluate treatment. Changes in clinical care may show a trend to normalization of red cell scatter characteristics, and therefore an improvement of red cell properties.

B.3 Acoustical scattering

An idealized "acoustical" material usually starts with the assumption that the stress can be represented as $\boldsymbol{\sigma} = -p\mathbf{1}$, where p is the pressure. For example, one may write, for small deformations in an inviscid, solid-like material, $p = -3\kappa \frac{\mathrm{tr}\nabla \boldsymbol{u}}{3}\mathbf{1}$, where $\boldsymbol{u}$ is the displacement and $\frac{\mathrm{tr}\nabla \boldsymbol{u}}{3}\mathbf{1}$ is the infinitesimal volumetric strain, with a corresponding strain energy of $W = \frac{1}{2}\frac{p^2}{\kappa}$.

B.3.1 Basic relations

By inserting the simplified expression of the stress $\boldsymbol{\sigma} = -p\mathbf{1}$ into the equation of equilibrium, we obtain

$$\nabla \cdot \boldsymbol{\sigma} = -\nabla p = \rho \ddot{\boldsymbol{u}}. \tag{B.6}$$

By taking the divergence of both sides, and recognizing that $\nabla \cdot \boldsymbol{u} = -\frac{p}{\kappa}$, where κ is the bulk modulus of the material, we obtain

$$\nabla^2 p = \frac{\rho}{\kappa}\ddot{p} = \frac{1}{c^2}\ddot{p}. \tag{B.7}$$

If we assume a harmonic solution, we obtain

$$p = Pe^{j(\boldsymbol{k}\cdot\boldsymbol{r}-\omega t)} \Rightarrow \dot{p} = Pj\omega e^{j(\boldsymbol{k}\cdot\boldsymbol{r}-\omega t)} \Rightarrow \ddot{p} = -P\omega^2 e^{j(\boldsymbol{k}\cdot\boldsymbol{r}-\omega t)} \tag{B.8}$$

and

$$\nabla p = Pj(k_x\boldsymbol{e}_x + k_y\boldsymbol{e}_y + k_z\boldsymbol{e}_z)e^{j(\boldsymbol{k}\cdot\boldsymbol{r}-\omega t)} \Rightarrow \nabla\cdot\nabla p = \nabla^2 p = -P\underbrace{(k_x^2 + k_y^2 + k_z^2)}_{||\boldsymbol{k}||^2} e^{j(\boldsymbol{k}\cdot\boldsymbol{r}-\omega t)}. \tag{B.9}$$

We insert these relations into Equation (B.7), and obtain an expression for the magnitude of the wave-number vector

$$-P||\boldsymbol{k}||^2 e^{j(\boldsymbol{k}\cdot\boldsymbol{r}-\omega t)} = -\frac{\rho}{\kappa}P\omega^2 e^{j(\boldsymbol{k}\cdot\boldsymbol{r}-\omega t)} \Rightarrow ||\boldsymbol{k}|| = \frac{\omega}{c}. \tag{B.10}$$

Equation (B.6) (balance of linear momentum) implies

$$\rho\ddot{\boldsymbol{u}} = -\nabla p = -Pj(k_x\boldsymbol{e}_x + k_y\boldsymbol{e}_y + k_z\boldsymbol{e}_z)e^{j(\boldsymbol{k}\cdot\boldsymbol{r}-\omega t)}. \tag{B.11}$$

Now we integrate once, which is equivalent to dividing by $-j\omega$, and obtain the velocity

$$\dot{\boldsymbol{u}} = \frac{Pj}{\rho\omega}(k_x\boldsymbol{e}_x + k_y\boldsymbol{e}_y + k_z\boldsymbol{e}_z)e^{j(\boldsymbol{k}\cdot\boldsymbol{r}-\omega t)}, \tag{B.12}$$

and do so again for the displacement

$$\boldsymbol{u} = \frac{Pj}{\rho\omega^2}(k_x\boldsymbol{e}_x + k_y\boldsymbol{e}_y + k_z\boldsymbol{e}_z)e^{j(\boldsymbol{k}\cdot\boldsymbol{r}-\omega t)}. \tag{B.13}$$

Thus, we have

$$||\dot{\boldsymbol{u}}|| = \frac{P}{c\rho}. \tag{B.14}$$

B.3.2 Reflection and ray-tracing

Now we turn to the problem of determining the p-wave scattering by large numbers of randomly distributed particles.

Ray-tracing

We consider cases where the particles are in the range of 10^{-4} m $\leq d \leq 10^{-3}$ m and the wavelengths are in the range of 10^{-6} m $\leq \lambda \leq 10^{-5}$ m. In such cases, geometric ray-tracing can be used to determine the amount of propagating incident energy that is reflected and the amount that is absorbed by multiple particles.

Incidence, reflection, and transmission

The reflection of a plane harmonic pressure wave at an interface is given by enforcing continuity of the (acoustical) pressure and disturbance velocity at that location; this yields the ratio between the incident and reflected pressures. We use a local coordinate system (Figure B.8) and require that the number of waves per unit length in the x direction be the same for the incident, reflected, and refracted (transmitted) waves, i.e.,

$$\boldsymbol{k}_i \cdot \boldsymbol{e}_x = \boldsymbol{k}_r \cdot \boldsymbol{e}_x = \boldsymbol{k}_t \cdot \boldsymbol{e}_x. \tag{B.15}$$

From the pressure balance at the interface, we have

$$P_i e^{j(\boldsymbol{k}_i\cdot\boldsymbol{r}-\omega t)} + P_r e^{j(\boldsymbol{k}_r\cdot\boldsymbol{r}-\omega t)} = P_t e^{j(\boldsymbol{k}_t\cdot\boldsymbol{r}-\omega t)}, \tag{B.16}$$

where P_i is the incident pressure ray, P_r is the reflected pressure ray, and P_t is the transmitted pressure ray. This forces a time-invariant relation to hold at all parts on the boundary, because the arguments of the exponential must be the same. This leads to ($k_i = k_r$)

$$k_i \sin\theta_i = k_r \sin\theta_r \Rightarrow \theta_i = \theta_r \tag{B.17}$$

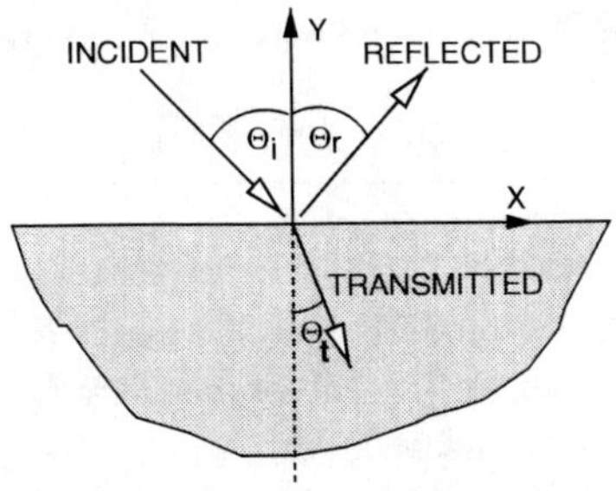

Figure B.8. *A local coordinate system for a ray reflection.*

and

$$k_i \sin\theta_i = k_t \sin\theta_t \Rightarrow \frac{k_i}{k_t} = \frac{\sin\theta_t}{\sin\theta_i} = \frac{\omega/c_t}{\omega/c_i} = \frac{c_i}{c_t} = \frac{v_i}{v_t} = \frac{n_t}{n_i}. \tag{B.18}$$

Equations (B.15) and (B.16) imply

$$P_i e^{j(\boldsymbol{k}_i\cdot\boldsymbol{r})} + P_r e^{j(\boldsymbol{k}_r\cdot\boldsymbol{r})} = P_t e^{j(\boldsymbol{k}_t\cdot\boldsymbol{r})}. \tag{B.19}$$

The continuity of the displacement, and hence the velocity

$$\boldsymbol{v}_i + \boldsymbol{v}_r = \boldsymbol{v}_t, \tag{B.20}$$

after use of Equation (B.14), leads to,

$$-\frac{P_i}{\rho_i c_i}\cos\theta_i + \frac{P_r}{\rho_r c_r}\cos\theta_r = -\frac{P_t}{\rho_t c_t}\cos\theta_t. \tag{B.21}$$

We solve for the ratio of the reflected and incident pressures to obtain

$$r = \frac{P_r}{P_i} = \frac{\hat{A}\cos\theta_i - \cos\theta_t}{\hat{A}\cos\theta_i + \cos\theta_t}, \tag{B.22}$$

where $\hat{A} \stackrel{\text{def}}{=} \frac{A_t}{A_i} = \frac{\rho_t c_t}{\rho_i c_i}$, ρ_t is the medium the ray encounters (transmitted), c_t is the corresponding sound speed in that medium, A_t is the corresponding acoustical impedance, ρ_i is the medium in which the ray was traveling (incident), c_i is the corresponding sound speed in that medium, and A_i is the corresponding acoustical impedance. The relationship (the law of refraction) between the incident and transmitted angles is $c_t \sin\theta_t = c_i \sin\theta_i$. Thus, we may write the Fresnel relation

$$r = \frac{\tilde{c}\hat{A}\cos\theta_i - (\tilde{c}^2 - \sin^2\theta_i)^{\frac{1}{2}}}{\tilde{c}\hat{A}\cos\theta_i + (\tilde{c}^2 - \sin^2\theta_i)^{\frac{1}{2}}}, \tag{B.23}$$

where $\tilde{c} \stackrel{\text{def}}{=} \frac{c_i}{c_t}$. The reflectance for the (acoustical) energy $R = r^2$ is

$$R = \left(\frac{P_r}{P_i}\right)^2 = \left(\frac{\hat{A}\cos\theta_i - \cos\theta_t}{\hat{A}\cos\theta_i + \cos\theta_t}\right)^2. \tag{B.24}$$

For the cases where $\sin\theta_t = \frac{\sin\theta_i}{\tilde{c}} > 1$, one may rewrite the reflection relation as

$$r = \frac{\tilde{c}\hat{A}\cos\theta_i - j(\sin^2\theta_i - \tilde{c}^2)^{\frac{1}{2}}}{\tilde{c}\hat{A}\cos\theta_i + j(\sin^2\theta_i - \tilde{c}^2)^{\frac{1}{2}}}, \tag{B.25}$$

where $j = \sqrt{-1}$. The reflectance is $R \stackrel{\text{def}}{=} r\bar{r} = 1$, where $\bar{r}$ is the complex conjugate. Thus, for angles above the critical angle $\theta_i \geq \theta_i^*$, all of the energy is reflected. We note that when $A_t = A_i$ and $c_i = c_t$, there is no reflection. Also, when $A_t \gg A_i$ or when $A_t \ll A_i$, $r \to 1$.

Remark. If one considers for a moment an incoming pressure wave (ray), which is incident on an interface between two general elastic media ($\mu \neq 0$), reflected shear waves must be generated in order to satisfy continuity of the traction, $[\![\boldsymbol{\sigma} \cdot \boldsymbol{n}]\!] = \mathbf{0}$. This is because

$$\left[\!\!\left[\left(3\kappa \operatorname{tr}\frac{\boldsymbol{\epsilon}}{3}\mathbf{1} + 2\mu\boldsymbol{\epsilon}'\right) \cdot \boldsymbol{n}\right]\!\!\right] = \mathbf{0}. \tag{B.26}$$

For an idealized acoustical medium, $\mu = 0$, no shear waves need to be generated to satisfy Equation (B.26).

Remark. Thus, in summary, the reflection of a plane harmonic pressure wave at an interface is given by enforcing continuity of the acoustical pressure and disturbance velocity at that location to yield the ratio between the incident and reflected pressures,

$$r = \frac{P_r}{P_i} = \frac{\hat{A}\cos\theta_i - \cos\theta_t}{\hat{A}\cos\theta_i + \cos\theta_t}, \tag{B.27}$$

where P_i is the incident pressure ray, P_r is the reflected pressure ray, $\hat{A} \stackrel{\text{def}}{=} \frac{\rho_t c_t}{\rho_i c_i}$, ρ_t is the medium the ray encounters (transmitted), c_t is the corresponding sound speed in that medium, ρ_i is the medium in which the ray was traveling (incident), and c_i is the corresponding sound speed in that medium. The relationship (the law of refraction) between the incident and transmitted angles is $c_t \sin\theta_t = c_i \sin\theta_i$. Thus, we may write

$$r = \frac{\tilde{c}\hat{A}\cos\theta_i - (\tilde{c}^2 - \sin^2\theta_i)^{\frac{1}{2}}}{\tilde{c}\hat{A}\cos\theta_i + (\tilde{c}^2 - \sin^2\theta_i)^{\frac{1}{2}}}, \tag{B.28}$$

where $\tilde{c} \stackrel{\text{def}}{=} \frac{c_i}{c_t}$. The reflectance for the acoustical energy is $R = r^2$. For the cases where $\sin\theta_t = \frac{\sin\theta_i}{\tilde{c}} > 1$, one may rewrite the reflection relation as

$$r = \frac{\tilde{c}\hat{A}\cos\theta_i - j(\sin^2\theta_i - \tilde{c}^2)^{\frac{1}{2}}}{\tilde{c}\hat{A}\cos\theta_i + j(\sin^2\theta_i - \tilde{c}^2)^{\frac{1}{2}}}, \tag{B.29}$$

where $j = \sqrt{-1}$. The reflectance is $R \stackrel{\text{def}}{=} r\bar{r} = 1$, where $\bar{r}$ is the complex conjugate. Thus, for angles above the critical angle $\theta_i \leq \theta_i^*$, all of the energy is reflected.

Bibliography

[1] Aboudi, J. 1991. *Mechanics of Composite Materials—A Unified Micromechanical Approach*. Elsevier.

[2] Ahmed, A. M. and Elghobashi, S. E. 2001. Direct numerical simulation of particle dispersion in homogeneous turbulent shear flows. *Physics of Fluids*, Vol. 13, 3346–3364.

[3] Ahmed, A. M. and Elghobashi, S. E. 2000. On the mechanisms of modifying the structure of turbulent homogeneous shear flows by dispersed particles. *Physics of Fluids*, Vol. 12, 2906–2930.

[4] Aitken, A. C. 1926. On Bernoulli's numerical solution of algebraic equations. *Proceedings of the Royal Society of Edinburgh*, Vol. 46, 289–305.

[5] Ames, W. F. 1977. *Numerical Methods for Partial Differential Equations*. 2nd edition. Academic Press.

[6] Arfken, G. 1970. *Mathematical Methods for Physicists*. 2nd edition. Academic Press.

[7] Armero, F. and Simo, J. C. 1992. A new unconditionally stable fractional step method for non-linear coupled thermomechanical problems. *International Journal for Numerical Methods in Engineering*, Vol. 35, 737–766.

[8] Armero, F. and Simo, J. C. 1993. A-priori stability estimates and unconditionally stable product formula algorithms for non-linear coupled thermoplasticity. *International Journal of Plasticity*, Vol. 9, 149–182.

[9] Armero, F. and Simo, J. C. 1996. Formulation of a new class of fractional-step methods for the incompressible MHD equations that retains the long-term dissipativity of the continuum dynamical system. In integration algorithms for classical mechanics. *The Fields Institute Communications*, Vol. 10, 1–23.

[10] Armero, F. 1999. Formulation and finite element implementation of a multiplicative model of coupled poro-plasticity at finite strains under fully saturated conditions. *Computer Methods in Applied Mechanics and Engineering*, Vol. 171, 205–241.

[11] Axelsson, O. 1994. *Iterative Solution Methods*. Cambridge University Press.

[12] Bale-Glickman, J., Selby, K., Saloner, D., and Savas, O. 2003. Physiological flow studies in exact-replica atherosclerotic carotid bifurcation. In *Proceedings of* 2003 *ASME International Mechanical Engineering Congress and Exposition*. Washington, D.C., 16–21.

[13] Bale-Glickman, J., Selby, K., Saloner, D., and Savas, O. 2003. Experimental flow studies in exact-replica phantoms of atherosclerotic carotid bifurcations under steady input conditions. *Journal of Biomechanical Engineering*, Vol. 125, 38–48.

[14] Barge, P. and Sommeria, J. 1995. Did planet formation begin inside persistent gaseous vortices? *Astronomy and Astrophysics*, Vol. 295, L1–4.

[15] Barranco, J., Marcus, P., and Umurhan, O. 2001. Scaling and asymptotics of coherent vortices in protoplanetary disks. In *Proceedings of the* 2000 *Summer Program—Center for Turbulence Research*. Stanford University Press, 85–96.

[16] Barranco, J. and Marcus, P. 2001. Vortices in protoplanetary disks and the formation of planetesimals. In *Proceedings of the* 2000 *Summer Program—Center for Turbulence Research*. Stanford University Press, 97–108.

[17] Barranco, J. and Marcus, P. 2005. Three-dimensional vortices in stratified protoplanetary disks. *Astrophysical Journal*, Vol. 623, 1157–1170.

[18] Bathe, K. J. 1996. *Finite Element Procedures*. Prentice–Hall.

[19] Becker, E. B., Carey, G. F., and Oden, J. T. 1980. *Finite Elements: An Introduction*. Prentice–Hall.

[20] Beckwith, S., Henning, T., and Nakagawa, Y. 2000. Dust particles in protoplanetary disks. In *Protostars & Planets* IV, Mannings, V., Boss, A. P., and Russell, S. S., editors, University of Arizona Press.

[21] Behringer, R. P. and Baxter, G. W. 1993. Pattern formation, complexity and time-dependence in granular flows. In *Granular matter—An interdisciplinary approach*. Mehta, A., editor, Springer-Verlag, 85–119.

[22] Behringer, R. P. 1993. The dynamics of flowing sand. *Nonlinear Science Today*, Vol. 3, 1.

[23] Behringer, R. P. and Miller, B. J. 1997. Stress fluctuations for sheared 3D granular materials. In *Proceedings, Powders & Grains* 97. Behringer, R. and Jenkins, J., editors. Balkema, 333–336.

[24] Behringer, R. P., Howell, D., and Veje, C. 1999. Fluctuations in granular flows. *Chaos*, Vol. 9, 559–572.

[25] Bender, J. and Fenton, R. 1970. On the flow capacity of automated highways. *Transport Science*, Vol. 4, 52–63.

[26] Benz, W. 2000. From dust to planets. *Spatium*, Vol. 6, 3–14.

[27] Benz, W. 1994. Impact simulations with fracture. 1. Method and Tests. *Icarus*, Vol. 107, 98–116.

[28] Berezin, Y. A., Hutter, K., and Spodareva, L. A. 1998. Stability properties of shallow granular flows. *International Journal of Non-Linear Mechanics*, Vol. 33, 647–658.

[29] Berger, S. A. and Jou, L. D. 2000. Flow in stenotic vessels. *Annual Review of Fluid Mechanics*, Vol. 32, 347–382.

[30] Berlyand, L. and Panchenko, A. 2007. Strong and weak blow-up of the viscous dissipation rates for concentrated suspensions. *To appear in the Journal of Fluid Mechanics.*

[31] Berlyand, L., Borcea, L., and Panchenko, A. 2005. Network approximation for effective viscosity of concentrated suspensions with complex geometries. *SIAM Journal of Mathematical Analysis*, Vol. 36, 1580–1628.

[32] Blum, J. and Wurm, G. 2000. Impact simulations on sticking, restructuring, and fragmentation of preplanetary dust aggregates. *Icarus*, Vol. 143, 138–146.

[33] Bohren, C. and Huffman, D. 1998. *Absorption and scattering of light by small particles.* Wiley Science Paperback Series.

[34] Bonabeau, E., Dorigo, M., and Theraulaz, G. 1999. *Swarm Intelligence: From Natural to Artificial Systems.* Oxford University Press.

[35] Born, M. and Wolf, E. 2003. *Principles of Optics.* 7th edition. Cambridge University Press.

[36] Breder, C. M. 1954. Equations descriptive of fish schools and other animal aggregations. *Ecology*, Vol. 35, 361–370.

[37] Brown, P. and Cooke, B. 2001. Model predictions for the 2001 Leonids and implications for Earth-orbiting satellites. *Monthly Notices of the Royal Astronomical Society*, Vol. 326, L19–L22.

[38] Chan, R. C., Chau, A. H., Karl, W. C., Nadkarni, S., Khalil, A. S., Shishkov, M., Tearney, G. J., Kaazempur-Mofrad, M. R., and Bouma, B. E. 2004. OCT-based arterial elastography: Robust estimation exploiting tissue biomechanics. *Optics Express*, Vol. 12, 4558–4572.

[39] Charalampopoulos, T. T. and Shu, G. 2003. Optical properties of combustion-synthesized iron oxide aggregates. *Applied Optics*, Vol. 42, 3957–3969.

[40] Charalampopoulos, T. T. and Shu, G. 2002. Effects of polydispersity of chainlike aggregates on light-scattering properties and data inversion. *Applied Optics*, Vol. 41, 723–733.

[41] Chau, A. H., Chan, R. C., Shishkov, M., MacNeill, B., Iftima, N., Tearney, G. J., Kamm, R. D., Bouma, B., and Kaazempur-Mofrad, M. R., 2004. Mechanical analysis of atherosclerotic plaques based on optical coherence tomography. *Annals of Biomedical Engineering*, Vol. 32, 1492–1501.

[42] Cho, H. and Barber, J. R. 1999. Stability of the three-dimensional Coulomb friction law. *Proceedings of the Royal Society*, Vol. 455, 839–862.

[43] Chokshi, A., Tielens, A. G. G. M., and Hollenbach, D. 1993. Dust coagulation. *The Astrophysical Journal*, Vol. 407, 806–819.

[44] Chow, C. Y. 1980. *An Introduction to Computational Fluid Dynamics*. Wiley.

[45] Chung, J., Grigoropoulos, C., and Greif, R. 1999. Infrared thermal velocimetry in MEMS-based fluid devices. *Journal of Microelectromechanical Systems*, Vol. 12, 365–371.

[46] Chyu, K. Y. and Shah, P. K. 2001. The role of inflammation in plaque disruption and thrombosis. In *Reviews in Cardiovascular Medicine*, Vol. 2, 82–91.

[47] Coleman, A. J. and Saunders, J. E. 1993. A review of the physical properties and biological effects of the high amplitude acoustic fields used in extracorporeal lithotripsy. *Ultrasonics*, Vol. 31, 75–89.

[48] Cuzzi, C. N., Dobrovolskis, A. R., and Champney, J. M. 1993. Particle-gas dynamics in the midplane of a protoplanetary nebula. *Icarus*, Vol. 106, 102–134.

[49] Dai, G., Kaazempur-Mofrad, M. R., Natarajan, S., Zhang, Y., Vaughn, S., Blackman, B. R., Kamm, R. D., Garcia-Cardena, G., and Gimbrone, M. A., Jr. 2004. Distinct endothelial phenotypes evoked by arterial waveforms derived from atherosclerosis-susceptible and resistant regions of human vasculature. *Proceedings of the National Academy of Sciences*, Vol. 101, 14871–14876.

[50] Davis, L. 1991. *Handbook of Genetic Algorithms*. Thompson Computer Press.

[51] Davies, M. J., Richardson, P. D., Woolf, N., Katz, D. R., and Mann, J. 1993. Risk of thrombosis in human atherosclerotic plaques: Role of extracellular lipid, macrophage, and smooth muscle cell content. *British Heart Journal*, Vol. 69, 377–381.

[52] Doltsinis, I. St. 1993. Coupled field problems—Solution techniques for sequential and parallel processing. In *Solving Large-Scale Problems in Mechanics*. Papadrakakis, M., editor. Wiley.

[53] Doltsinis, I. St. 1997. Solution of coupled systems by distinct operators. *Engineering Computations*, Vol. 14, 829–868.

[54] Dominik, C. and Tielens, A. G. G. M. 1997. The physics of dust coagulation and the structure of dust aggregates in space. *The Astrophysical Journal*, Vol. 480, 647–673.

[55] Donev, A., Cisse, I., Sachs, D., Variano, E. A., Stillinger, F., Connelly, R., Torquato, S., and Chaikin, P. 2004. Improving the density of jammed disordered packings using ellipsoids. *Science*, Vol. 303, 990–993.

[56] Donev, A., Stillinger, F. H., Chaikin, P. M., and Torquato, S. 2004. Unusually dense crystal ellipsoid packings. *Physical Review Letters*, Vol. 92, 255506.

[57] Donev, A., Torquato S., and Stillinger, F. 2005. Neighbor list collision-driven molecular dynamics simulation for nonspherical hard particles—I. Algorithmic details. *Journal of Computational Physics*, Vol. 202, 737.

[58] Donev, A., Torquato, S., and Stillinger, F. 2005. Neighbor list collision-driven molecular dynamics simulation for nonspherical hard particles—II. Application to ellipses and ellipsoids. *Journal of Computational Physics*, Vol. 202, 765.

[59] Donev, A., Torquato, S., and Stillinger, F. H. 2005. Pair correlation function characteristics of nearly jammed disordered and ordered hard-sphere packings. *Physical Review* E, Vol. 71, 011105.

[60] Druzhinin, O. A. and Elghobashi, S. E. 2001. Direct numerical simulation of a three-dimensional spatially-developing bubble-laden mixing layer with two-way coupling. *Journal of Fluid Mechanics*, Vol. 429, 23–61.

[61] Duran, J. 1997. *Sands, Powders and Grains. An Introduction to the Physics of Granular Matter.* Springer-Verlag.

[62] Eber, S. and Lux, S. E. 2004. Hereditary spherocytosis—Defects in proteins that connect the membrane skeleton to the lipid bilayer. *Seminars in Hematology*, Vol. 41, 118–141.

[63] Elmore, W. C. and Heald, M. A. 1985. *Physics of Waves*. Dover.

[64] Evans, E. A. and Fung, Y. C. 1972. Improved measurements of the erythrocyte geometry. *Microvascular Research*, Vol. 4, 335.

[65] Farhat, C., Lesoinne, M., and Maman, N. 1995. Mixed explicit/implicit time integration of coupled aeroelastic problems: Three-field formulation, geometric conservation and distributed solution. *International Journal for Numerical Methods in Fluids*, Vol. 21, 807–835.

[66] Farhat, C. and Lesoinne, M. 2000. Two efficient staggered procedures for the serial and parallel solution of three-dimensional nonlinear transient aeroelastic problems. *Computer Methods in Applied Mechanics and Engineering*, Vol. 182, 499–516.

[67] Farhat, C., van der Zee, G., and Geuzaine, P. 2006. Provably second-order time-accurate loosely-coupled solution algorithms for transient nonlinear computational aeroelasticity. *Computer Methods in Applied Mechanics and Engineering*, Vol. 195, 1973–2001.

[68] Ferrante, A. and Elghobashi, S. E. 2003. On the physical mechanisms of two-way coupling in particle-laden isotropic turbulence. *Physics of Fluids*, Vol. 15, 315–329.

[69] Forget, B. G. and Cohen, A. R. 2005. Thalassemia syndromes. In *Hematology: Basic Principles and Practice*, 4th edition, Hoffman B. E., ed. Elsevier.

[70] Fowles, G. R. 1989. *Introduction to Modern Optics*, 2nd edition. Dover, reissue.

[71] Frenklach, M. and Carmer, C. S. 1999. Molecular dynamics using combined quantum and empirical forces: Application to surface reactions. *Advances in Classical Trajectory Methods.* Vol. 4, 27–63.

[72] Fuster, V. 2002. *Assessing and modifying the vulnerable atherosclerotic plaque*. Futura Publishing.

[73] Gallagher, P. G. 2004. Hereditary elliptocytosis: Spectrin and protein 4.1R. *Seminars in Hematology*, Vol. 41, 142–164.

[74] Gallagher, P. G. 2004. Update on the clinical spectrum and genetics of red blood cell membrane disorders. *Current Hematology Reports*, Vol. 3, 85–91.

[75] Gazi, V. and Passino, K. M. 2002. Stability analysis of swarms. In *Proceedings of the American Control Conference*, Anchorage, AK.

[76] Gill, P., Murray, W., and Wright, M. 1995. *Practical Optimization*. Academic Press.

[77] Goldberg, D. E. 1989. *Genetic Algorithms in Search, Optimization and Machine Learning*. Addison-Wesley.

[78] Goldberg, D. E. and Deb, K. 2000. Special issue on genetic algorithms. *Computer Methods in Applied Mechanics and Engineering*, Vol. 186 (2-4), 121–124.

[79] Goldsmith, W. 2001. *Impact: The Theory and Physical Behavior of Colliding Solids.* Dover.

[80] Gray, J. M. N. T, Wieland, M., and Hutter, K. 1999. Gravity-driven free surface flow of granular avalanches over complex basal topography. *Proceedings of the Royal Society of London, Series* A, Vol. 455, 1841–1874.

[81] Gray, J. M. N. T. and Hutter, K. 1997. Pattern formation in granular avalanches. *Continuum Mechanics and Thermodynamics*, Vol. 9, 341–345.

[82] Gray, J. M. N. T. 2001. Granular flow in partially filled slowly rotating drums. *Journal of Fluid Mechanics*, Vol. 441, 1–29.

[83] Grazier, K. R., Newman, W. I., Kaula, W. M., and Hyman, J. M. 2000. Dynamical evolution of planetesimals in the outer solar system I. The Jupiter/Saturn zone. *Icarus*, Vol. 140, 341–352.

[84] Grazier, K. R., Newman, W. I., Varadi, F., Kaula, W. M., and Hyman, J. M. 1999. Dynamical evolution of planetesimals in the outer solar system II. The Saturn/Uranus and Uranus/Neptune zones. *Icarus*, Vol. 140, 353–368.

[85] Greve, R. and Hutter, K. 1993. Motion of a granular avalanche in a convex and concave curved chute: Experiments and theoretical predictions. *Philosophical Transactions of the Royal Society of London, Series* A, Vol. 342, 573–600.

[86] Gross, H. 2005. *Handbook of Optical Systems. Fundamentals of Technical Optics*, Gross, H., editor. Wiley-VCH.

[87] Haile, J. M. 1992. *Molecular Dynamics Simulations: Elementary Methods*. Wiley.

[88] Hale, J. & Kocak, H. 1991. *Dynamics and Bifurcations*. Springer-Verlag.

[89] Hase, W. L. 1999. *Molecular Dynamics of Clusters, Surfaces, Liquids, & Interfaces. Advances in Classical Trajectory Methods*, Vol. 4. JAI Press.

[90] Hashin, Z. 1983. Analysis of composite materials: A survey. *ASME Journal of Applied Mechanics*, Vol. 50, 481–505.

[91] Hecht, E. 2002. *Optics*. 4th edition. Addison–Wesley.

[92] Hedrick, J. K. and Swaroop, D. 1993. Dynamic coupling in vehicles under automatic control. In *Proceedings of the 13th IAVSD Symposium*, 209–220.

[93] Hedrick, J. K., Tomizuka, M., and Varaiya, P. 1994. Control issues in automated highway systems. *IEEE Controls Systems Magazine*, Vol. 14 (6), 21–32.

[94] Holland, J. H. 1975. *Adaptation in Natural and Artificial Systems*. University of Michigan Press.

[95] Hughes, T. J. R. 1989. *The Finite Element Method*. Prentice–Hall.

[96] Hutter, K. 1996. Avalanche dynamics. In *Hydrology of Disasters*, Singh, V. P., editor. Kluwer Academic Publishers, 317–394.

[97] Hutter, K., Koch, T., Plüss, C., and Savage, S. B. 1995. The dynamics of avalanches of granular materials from initiation to runout. Part II. Experiments. *Acta Mechanica*, Vol. 109, 127–165.

[98] Hutter, K. and Rajagopal, K. R. 1994. On flows of granular materials. *Continuum Mechanics and Thermodynamics*, Vol. 6, 81–139.

[99] Hutter, K., Siegel, M., Savage, S. B., and Nohguchi, Y. 1993. Two-dimensional spreading of a granular avalanche down an inclined plane. Part I: Theory. *Acta Mechanica*, Vol. 100, 37–68.

[100] Jaeger, H. M. and Nagel, S. R. 1992. La Physique de l'Etat Granulaire. *La Recherche*, Vol. 249, 1380.

[101] Jaeger, H. M. and Nagel, S. R. 1992. Physics of the granular state. *Science*, Vol. 255, 1523.

[102] Jaeger, H. M. and Nagel, S. R. 1993. La Fisica del Estado Granular. *Mundo Cientifico*, Vol. 132, 108.

[103] Jaeger, H. M., Knight, J. B., Liu, C. H., and Nagel, S. R. 1994. What is shaking in the sand box? *Materials Research Society Bulletin*, Vol. 19, 25.

[104] Jaeger, H. M., Nagel, S. R., and Behringer, R. P. 1996. The physics of granular materials. *Physics Today*, Vol. 4, 32.

[105] Jaeger, H, M., Nagel, S. R., and Behringer, R. P. 1996. Granular solids, liquids and gases, *Reviews of Modern Physics*, Vol. 68, 1259.

[106] Jaeger, H. M. and Nagel, S. R. 1997. Dynamics of granular material. *American Scientist*, Vol. 85, 540.

[107] Jenkins, J. T. and Strack, O. D. L. 1993. Mean-field inelastic behavior of random arrays of identical spheres. *Mechanics of Materials*, Vol. 16, 25–33.

[108] Jenkins, J. T. and La Ragione, L. 1999. Particle spin in anisotropic granular materials. *The International Journal of Solids Structures*, Vol. 38, 1063–1069.

[109] Jenkins, J. T. and Koenders, M. A. 2004. The incremental response of random aggregates of identical round particles. *The European Physics Journal* E—*Soft Matter*, Vol. 13, 113–123.

[110] Jenkins, J. T., Johnson, D., La Ragione, L., and Makse, H. 2005. Fluctuations and the effective moduli of an isotropic, random aggregate of identical, frictionless spheres. *The Journal of Mechanics and Physics of Solids*, Vol. 53, 197–225.

[111] Johnson, K. 1985. *Contact Mechanics.* Cambridge University Press.

[112] Jou, L. D. and Berger, S. A. 1998. Numerical simulation of the flow in the carotid bifurcation. *Theoretical and Computational Fluid Mechanics*, Vol. 10, 239–248.

[113] Kaazempur-Mofrad, M. R. and Ethier, C. R. 2001. Mass transport in an anatomically realistic human right coronary artery. *Annals of Biomedical Engineering*, Vol. 29(2), 121–127.

[114] Kaazempur-Mofrad, M. R., Younis, H. F., Patel, S., Isasi, A. G., Chung, C., Chan, R. C., Hinton, D. P., Lee, R. T., and Kamm, R. D. 2003. Cyclic strain in human carotid bifurcation and its potential correlation to atherogenesis: Idealized and anatomically-tealistic models. *Journal of Engineering Mathematics*, Vol. 47(3-4), 299–314.

[115] Kaazempur-Mofrad, M. R., Younis, H. F., Isasi, A. G., Chan, R. C., Hinton, D. P., Sukhova, G., LaMuraglia, G. M., Lee, R. T., and Kamm, R. D. 2004. Characterization of the atherosclerotic carotid bifurcation using MRI, finite element modeling and histology. *Annals of Biomedical Engineering*, Vol. 32, 932–946.

[116] Kahn, S. R., Finlayson, B., and Hackett, R. L. 1983. Stone matrix as proteins adsorbed on crystal surfaces: A microscopic study. *Scanning Electron Microscopy*, Vol. 1983, 379–385.

[117] Kahn, S. R. and Hackett, R. L. 1993. Role of organic matrix in urinary stone formation: An ultrastructural study of crystal matrix interface of calcium oxalate monohydrate stones. *Journal of Urology*, Vol. 150, 239–245.

[118] Kahn, S. R. and Hackett. R. L. 1984. Microstructure of decalcified human calcium oxalate urinary stones. *Scanning Electron Microscopy*, Vol. II, 935–941.

[119] Kansaal, A., Torquato, S., and Stillinger, F. 2002. Diversity of order and densities in jammed hard-particle packings. *Physical Review* E, Vol. 66, 041109.

[120] Kennedy, J. and Eberhart, R. 2001. *Swarm Intelligence.* Morgan Kaufmann Publishers.

[121] Khalil, A. S., Chan, R. C., Chau, A. H., Bouma, B. E., and Kaazempur-Mofrad, M. R. 2005. Tissue elasticity estimation with optical coherence elastography: Toward mechanical characterization of in vivo soft tissue. *Annals of Biomedical Engineering*, Vol. 33, 1631–1639.

[122] Khalil, A. S., Bouma, B. E., and Kaazempur-Mofrad, M. R. 2006. A combined FEM/genetic algorithm for vascular soft tissue elasticity estimation. Cardiovascular Engineering, Vol. 6(3), 93–102.

[123] Kikuchi, N. and Oden, J. T. 1988. *Contact Problems in Elasticity: A Study of Variational Inequalities and Finite Element Methods.* SIAM.

[124] Kim, K. M. 1982. The stones. *Scanning Electron Microscopy*, Vol. IV, 1635–1660.

[125] Klarbring, A. 1990. Examples of non-uniqueness and non-existence of solutions to quasistatic contact problems with friction. *Ingenieur-Archiv*, Vol. 60, 529–541.

[126] Koch, T., Greve, R., and Hutter, K. 1994. Unconfined flow of granular avalanches along a partly curved surface. II. Experiments and numerical computations. *Proceedings of the Royal Society of London, Series* A, Vol. 445, 415–435.

[127] Kokubo, E. and Ida, S. 2000. Formation of protoplanets from planetesimals in the solar nebula. *Icarus*, Vol. 143, 15–270.

[128] Kokubo, E. and Ida, S. 1996. On runaway growth of planetesimals. *Icarus*, Vol. 123, 180–191.

[129] Lagaros, N., Papadrakakis, M., and Kokossalakis, G. 2002. Structural optimization using evolutionary algorithms. *Computers and Structures*, Vol. 80, 571–589.

[130] Lesoinne, M. and Farhat, C. 1998. Free staggered algorithm for nonlinear transient aeroelastic problems. *AIAA Journal*, Vol. 36, 1754–1756.

[131] Le Tallec, P. and Mouro, J. 2001. Fluid structure interaction with large structural displacements. *Computer Methods in Applied Mechanics and Engineering*, Vol. 190, 3039–3067.

[132] Lewis, R. W. and Schrefler, B. A. 1998. *The Finite Element Method in the Static and Dynamic Deformation and Consolidation of Porous Media*, 2nd edition. Wiley Press.

[133] Lewis, R. W., Schrefler, B. A., and Simoni, L. 1992. Coupling versus uncoupling in soil consolidation. *International Journal for Numerical and Analytical Methods in Geomechanics*, Vol. 15, 533–548.

[134] Libby, P. 2001. Current concepts of the pathogenesis of the acute coronary syndromes. *Circulation*, Vol. 104, 365–372.

[135] Libby, P. 2001. The vascular biology of atherosclerosis. In *Heart Disease. A Textbook of Cardiovascular Medicine*, Braunwald, E., Zipes, D. P., and Libby, P., editors, 6th edition. W. B. Saunders Company, Chapter 30, 995–1009.

[136] Libby, P., Ridker, P. M., and Maseri, A. 2002. Inflammation and atherosclerosis. *Circulation*, Vol. 105, 1135–1143.

[137] Libby, P. and Aikawa, M. 2002. Stabilization of atherosclerotic plaques: New mechanisms and clinical targets. *Nature Medicine*, Vol. 8, 1257–1262.

[138] Lissauer, J. J. 1993. Planet formation. *Annual Review of Astronomy and Astrophysics*, Vol. 31, 129–174.

[139] Liu, C. H., Jaeger, H. M., and Nagel, S. R. 1991. Finite size effects in a sandpile. *Physical Review* A, Vol. 43, 7091.

[140] Liu, C. H. and Nagel, S. R. 1993. Sound in a granular material: Disorder and Nonlinearity. *Physical Review* B, Vol. 48, 15646.

[141] Loree, H. M., Kamm, R. D., Stringfellow, R. G., and Lee, R. T. 1992. Effects of fibrous cap thickness on peak circumferential stress in model atherosclerotic vessels. *Circulation Research*, Vol. 71, 850–858.

[142] Luenberger, D. 1974. *Introduction to Linear and Nonlinear Programming.* Addison-Wesley.

[143] Luo, L. and Dornfeld, D. A. 2001. Material removal mechanism in chemical mechanical polishing: Theory and modeling. *IEEE Transactions: Semiconductor Manufacturing*, Vol. 14, 112–133.

[144] Luo, L. and Dornfeld, D. A. 2003. Effects of abrasive size distribution in chemical-mechanical planarization: Modeling and verification. *IEEE Transactions: Semiconductor Manufacturing*, Vol. 16, 469–476.

[145] Luo, L. and Dornfeld, D. A. 2003. Material removal regions in chemical mechanical planarization for sub-micron integrated circuit fabrication: Coupling effects of slurry chemicals, abrasive size distribution, and wafer-pad contact area. *IEEE Transactions: Semiconductor Manufacturing*, Vol. 16, 45–56.

[146] Luo, L. and Dornfeld, D. A. 2002. Optimization of CMP from the viewpoint of consumable effects. *Journal of the Electrochemical Society*, Vol. 150, G807–G815.

[147] Martins, J. A. C. and Oden, J. T. 1987. Existence and uniqueness results in dynamics contact problems with nonlinear normal and friction interfaces. *Nonlinear Analysis*, Vol. 11, 407–428.

[148] Mitchell, P. and Frenklach, M. 2003. Particle aggregation with simultaneous surface growth. *Physical Review* E, Vol. 67, 061407.

[149] Moelwyn-Hughes, E. A. 1961. *Physical Chemistry*. Pergamon.

[150] Mura, T. 1993. *Micromechanics of Defects in Solids*. 2nd edition. Kluwer Academic Publishers.

[151] Nagel, S. R. 1992. Instabilities in a Sandpile, *Review of Modern Physics*, Vol. 64, 321.

[152] Nemat-Nasser, S. and Hori, M. 1999. *Micromechanics: Overall Properties of Heterogeneous Solids*. 2nd edition. Elsevier.

[153] Nye, J. F. 1957. *Physical Properties of Crystals*. Oxford University Press.

[154] Oden, J. T. and Pires, E. 1983. Nonlocal and nonlinear friction laws and variational principles for contact problems in elasticity. *ASME Journal of Applied Mechanics*, Vol. 50, 67–76.

[155] Onwubiko, C. 2000. *Introduction to Engineering Design Optimization*. Prentice–Hall.

[156] Papadrakakis, M., Lagaros, N., Thierauf, G., and Cai, J. 1998. Advanced solution methods in structural optimisation using evolution strategies. *Engineering Computational Journal*, Vol. 15, 12–34.

[157] Papadrakakis, M., Lagaros, N., and Tsompanakis, Y. 1998. Structural optimization using evolution strategies and neutral networks. *Computer Methods in Applied Mechanics and Engineering*, Vol. 156, 309–335.

[158] Papadrakakis, M., Lagaros, N., and Tsompanakis, Y. 1999. Optimization of large-scale 3D trusses using evolution strategies and neural networks. *International Journal of Space Structures*, Vol. 14, 211–223.

[159] Papadrakakis, M., Tsompanakis, J., and Lagaros, N. 1999. Structural shape optimisation using evolution strategies. *Engineering Optimization*, Vol. 31, 515–540.

[160] Papadrakakis, M., Lagaros, N., Tsompanakis, Y., and Plevris, V. 2001. Large scale structural optimization: Computational methods and optimization algorithms. *Archives of Computational Methods in Engineering, State of the Art Reviews*, Vol. 8, 239–301.

[161] Park, K. C. and Felippa, C. A. 1983. Partitioned analysis of coupled systems. In *Computational Methods for Transient Analysis*. Belytschko, T. and Hughes, T. J. R., editors. North-Holland.

[162] Piperno, S. 1997. Explicit/implicit fluid/structure staggered procedures with a structural predictor and fluid subcycling for 2D inviscid aeroelastic simulations. *International Journal of Numerical Methods in Fluids*, Vol. 25, 1207–1226.

[163] Piperno, S., Farhat, C., and Larrouturou, B. 1995. Partitioned procedures for the transient solution of coupled aeroelastic problems—Part I: Model problem, theory, and two-dimensional application. *Computer Methods in Applied Mechanics and Engineering*, Vol. 124, 79–112.

[164] Piperno, S. and Farhat, C. 2001. Partitioned procedures for the transient solution of coupled aeroelastic problems—Part II: Energy transfer analysis and three-dimensional applications. *Computer Methods in Applied Mechanics and Engineering*, Vol. 190, 3147–3170.

[165] Pittomvils, G., Vanduersen, H., Wevers, M., Lafaut, J. P., De Ridder, D., De Meester, P., Boving, R., and Baert, L. 1994. The influence of internal stone structure on the fracture behavior of urinary calculi. *Ultrasound in Medicine and Biology*, Vol. 20, 803–810.

[166] Pollack, J. N., Hollenbach, D., Beckwith, S., Simonelli, D. P., Roush, T., and Fong, W. 1994. Composition and radiative properties in molecular clouds and accretion disks. *The Astrophysical Journal*, Vol. 421, 615–639.

[167] Pöschel, T. and Schwager, T. 2004. *Computational Granular Dynamics*. Springer-Verlag.

[168] Rapaport, D. C. 1995. *The Art of Molecular Dynamics Simulation*. Cambridge University Press.

[169] Richardson, P. D., Davies, M. J., and Born, G. V. R. 1989. Influence of plaque configuration and stress distribution on fissuring of coronary atherosclerotic plaques. *Lancet*, Vol. 2 (8669), 941–944.

[170] Rietema, K. 1991. *Dynamics of Fine Powders*. Springer-Verlag.

[171] Schlick, T. 2000. *Molecular Modeling and Simulation. An Interdisciplinary Guide*. Springer-Verlag.

[172] Schmidt, L. 1998. *The Engineering of Chemical Reactions*. Oxford University Press.

[173] Schrefler, B. A. 1985. A partitioned solution procedure for geothermal reservoir analysis. *Communications in Applied Numerical Methods*, Vol. 1, 53–56.

[174] Shah, P. K. 1997. Plaque disruption and coronary thrombosis: New insight into pathogenesis and prevention. *Clinical Cardiology*, Vol. 20 (Suppl. II), II-38–II-44.

[175] Shamma, J. S. 2001. A connection between structured uncertainty and decentralized control of spatially invariant systems. In *Proceedings of the* 2001 *American Control Conference*, Vol. 4, 3117–3121.

[176] Shanks, D. 1955. Non-linear transformations of divergent or slowly convergent sequences. *Journal of Mathematics and Physics*, Vol. XXXIV, 1–42.

[177] Shin, Y., Chung, J., Kladias, N., Panides, E., Domoto, J., and Grigoropoulos, C. P. 2005. Compressible flow of liquid in standing wave tube. *Journal of Fluid Mechanics*, Vol. 536, 321–345.

[178] Steinberg, M. H., Benz, E. J., Jr., Adewoye, H. A., and Hoffman, B. E. 2005. Pathobiology of the human erythrocyte and its hemoglobins. In *Hematology: Basic principles and practice*, Hoffman, B. E., editor, 4th edition. Elsevier.

[179] Stillinger, F. H. and Weber, T. A. 1985. Computer simulation of local order in condensed phases of silicon. *Physical Review* B, Vol. 31, 5262–5271.

[180] Stroud, J. S., Berger, S. A., and Saloner, D. 2002. Numerical analysis of flow through a severely stenotic carotid artery bifurcation. *Journal of Biomedical Engineering*, Vol. 124, 9–21.

[181] Stroud, J. S., Berger, S. A., and Saloner, D. 2000. Influence of stenosis morphology on flow through severely stenotic vessels: Implications of plaque rupture. *Journal of Biomechanics*, Vol. 33, 443–455.

[182] Supulver, K. D. and Lin, D. N. C. 2000. Formation of icy planetesimals in a turbulent solar nebula. *Icarus*, Vol. 146, 525–540

[183] Swaroop, D. and Hedrick, J. K. 1996. String stability of interconnected systems. *IEEE Transactions on Automatic Control*, Vol. 41, 349–356.

[184] Swaroop, D. and Hedrick, J. K. 1999. Constant spacing strategies for platooning in automated highway systems. *Journal ofDynamic Systems, Measurement and Control*, Vol. 121, 462–470.

[185] Szabo, B. and Babúska, I. 1991. *Finite Element Analysis*. Wiley Interscience.

[186] Tabor, D. 1975. Interaction between surfaces: Adhesion and friction. In *Surface Physics of Materials*, Blakely, J. M., editor, Vol. II, Academic Press, Chapter 10.

[187] Taflove, A. and Hagness, S. 2005. *Computational Electromagnetics*, 3rd edition. Artech House, Inc.

[188] Tai, Y.-C., Noelle, S., Gray, J. M. N. T., and Hutter, K. 2002. Shock capturing and front tracking methods for granular avalanches. *Journal of Computational Physics*, Vol. 175, 269–301.

[189] Tai, Y.-C., Gray, J. M. N. T., Hutter, K., and Noelle, S. 2001. Flow of dense avalanches past obstructions. *Annals of Glaciology*, Vol. 32, 281–284.

[190] Tai, Y.-C., Noelle, S., Gray, J. M. N. T., and Hutter, K. 2001. An accurate shock-capturing finite-difference method to solve the Savage-Hutter equations in avalanche dynamics. *Annals of Glaciology*, Vol. 32, 263–267.

[191] Tanga, P., Babiano, A., Dubrulle, B., and Provenzale, A. 1996. Forming planetesimals in vortices. *Icarus*, Vol. 121, pp. 158–170.

[192] Telford, W. M., Geldart, L. P., and Sheriff, R. E. 1990. *Applied Geophysics*, 2nd edition. Cambridge University Press.

[193] Tersoff, J. 1988. Empirical interatomic potential for carbon, with applications to amorphous carbon. *Physical Review Letters*, Vol. 61, 2879–2882.

[194] Torquato, S. 2002. *Random Heterogeneous Materials: Microstructure and Macroscopic Properties*. Springer-Verlag.

[195] van de Hulst, H. C. 1981. *Light Scattering by Small Particles*. Dover.

[196] Tuzun, U. and Walton, O. R. 1992. Micro-mechanical modelling of load dependent friction in contacts of elastic spheres. *Journal of Physics* D: *Applied Physics*, Vol. 25, A44–A52.

[197] van der Wal, A. C. and Becker, A. E. 1999. Atherosclerotic plaque rupture—Pathologic basis of plaque stability and instability. Cardiovascular Research, Vol. 41, 334–344.

[198] Weidenschilling, S. J. and Cuzzi, J. N. 1993. Formation of planetesimals in the solar nebula. In *Protostars and Planets* III. Levy, E. H. and Lunine, J. I., editors. University of Arizona Press, 1031–1060.

[199] Weidenschilling, S. J., Spaute, D., Davis, D. R., Marzari, F., and Ohtsuki, K. 1997. Accretional evolution of a planetesimal swarm. *Icarus*, Vol. 128, 429–455.

[200] Widom, B. 1966. Random sequential addition of hard spheres to a volume. *Journal of Chemical Physics*, Vol. 44, 3888–3894.

[201] Wieland, M., Gray, J. M. N. T., and Hutter, K. 1999. Channelized free-surface flow of cohesionless granular avalanches in a chute with shallow lateral curvature. *Journal of Fluid Mechanics*, Vol. 392, 73–100.

[202] Williamson, S. D., Lam, Y., Younis, H. F., Huang, H., Patel, S., Kaazempur-Mofrad, M. R., and Kamm, R. D. 2003. On the sensitivity of wall stresses in diseased arteries to variable material properties. *Journal of Biomechanical Engineering*, Vol. 125, 147–155.

[203] Wriggers, P. 2002. *Computational Contact Mechanics*. John Wiley.

[204] Wurm, G., Blum, J., and Colwell, J. E. 2001. A new mechanism relevant to the formation of planetesimals in the solar nebula. *Icarus*, Vol. 151, 318–321.

[205] Younis, H. F., Kaazempur-Mofrad, M. R., Chan, R., Chau, A. H., Hinton, D. P., Isasi, A. G., Kim, L. A, and Kamm, R. D. 2004. Hemodynamics and wall mechanics in human carotid bifurcation and its consequences for atherogensis: Investigation of inter-individual variation. *Biomechanics and Modeling in Mechanobiology*, Vol. 3, 17–32.

[206] Zienkiewicz, O. C. 1984. Coupled problems and their numerical solution. In *Numerical Methods in Coupled Systems*, Lewis, R. W., Bettes, P., and Hinton, E., editors. Wiley, 35–58.

[207] Zienkiewicz, O. C. and Taylor R. L. 1991. *The Finite Element Method*. Vols. I and II. McGraw-Hill.

[208] Zohdi, T. I. 2002. An adaptive-recursive staggering strategy for simulating multi-field coupled processes in microheterogeneous solids. *The International Journal of Numerical Methods in Engineering*, Vol. 53, 1511–1532.

[209] Zohdi, T. I. 2003. Computational design of swarms. *The International Journal of Numerical Methods in Engineering*, Vol. 57, 2205–2219.

[210] Zohdi, T. I. 2004. Modeling and simulation of a class of coupled thermo-chemo-mechanical processes in multiphase solids. *Computer Methods in Applied Mechanics and Engineering*, Vol. 193, 679–699.

[211] Zohdi, T. I. 2003. Genetic optimization of statistically uncertain microheterogeneous solids. *Philosophical Transactions of the Royal Society: Mathematical, Physical and Engineering Sciences*, Vol. 361, 1021–1043.

[212] Zohdi, T. I. 2004. Modeling and direct simulation of near-field granular flows. *The International Journal of Solids and Structures*, Vol. 42, 539–564.

[213] Zohdi, T. I. 2004. A computational framework for agglomeration in thermo-chemically reacting granular flows. *Proceedings of the Royal Society*, Vol. 460, 3421–3445.

[214] Zohdi, T. I., Monteiro, P. J. M., and Lamour, V. 2002. Extraction of elastic moduli from granular compacts. *The International Journal of Fracture/Letters in Micromechanics*, Vol. 115, L49–L54.

[215] Zohdi, T. I. 2003. Constrained inverse formulations in random material design. *Computer Methods in Applied Mechanics and Engineering*, Vol. 192, 3179–3194.

[216] Zohdi, T. I. and Wriggers, P. 2005. *Introduction to Computational Micromechanics*. Springer-Verlag.

[217] Zohdi, T. I. 2005. Charge-induced clustering in multifield granular flow. *The International Journal of Numerical Methods in Engineering*, Vol. 62, 870–898.

[218] Zohdi, T. I. 2006. Computation of the coupled thermo-optical scattering properties of random particulate systems. *Computer Methods in Applied Mechanics and Engineering*, Vol. 195, 5813–5830.

[219] Zohdi, T. I. 2006. On the optical thickness of disordered particulate media. *Mechanics of Materials*, Vol. 38, 969–981.

[220] Zohdi, T. I., Holzapfel, G. A., and Berger, S. A. 2004. A phenomenological model for atherosclerotic plaque growth and rupture. *The Journal of Theoretical Biology*, Vol. 227, 437–443.

[221] Zohdi, T. I. 2005. A simple model for shear stress mediated lumen reduction in blood vessels. *Biomechanics and Modeling in Mechanobiology*, Vol. 4, 57–61.

[222] Zohdi, T. I. and Szeri, A. J. 2005. Fatigue of kidney stones with heterogeneous microstructure subjected to shock wave lithotripsy. *Journal of Biomedical Materials Research: Part* B—*Applied Biomaterials*, Vol. 75B, 351–358.

[223] Zohdi, T. I and Kuypers, F. A. 2006. Modeling and rapid simulation of multiple red blood cell light scattering *Journal of the Royal Society Interface*, Vol. 3, 823–831.

[224] Zohdi, T. I., in press. Computation of strongly coupled multifield interaction in particle-fluid systems. *Computer Methods in Applied Mechanics and Engineering.* Available online at the journal website, http://www.elsevier.com

[225] Zohdi, T. I., submitted. Particle collision and adhesion under the influence of near-fields. Preprint available upon request. *Journal of Mechanics of Materials and Structures*, http://pjm.math.berkeley.edu/jomms/2006/1-1/po1.xhtml

Index